Student Solutions Manual to Accompany

FOURTH EDITION

Elementary Statistics

Mario F. Triola

by Donald K. Mason

ISBN 0-8053-0273-5

7 8 9 10 -AL- 95 94 93 92 91

The Benjamin/Cummings Publishing Company, Inc.
390 Bridge Parkway
Redwood City, California 94065

PREFACE

This manual contains the solutions to most of the odd problems of the textbook, Elementary Statistics, Fourth Edition, by Mario Triola. In the worked problems, the detail of the computations is provided along with reasons for making the various steps. By this means students can follow the reasoning used to arrive at a solution. In addition, reference to formulas and examples of the text should assist the student.

In many instances the answer in this manual will differ slightly from the answer provided in the text. This is, for the most part, due to differences in the selection of significant digits. For example, if you are dividing by the square root of 3, you will get a different answer if you divide by 1.7 as opposed to dividing by 1.732. Both methods might be correct, but the numerical answers may differ slightly.

The problems of Chapter 1 were not worked. In addition some problems through the other chapters were omitted. In most cases the reason for the omission was that it was felt that there had been a sufficient number of similar problems worked. Consequently, there would not be sufficient assistance to students to warrant inclusion.

I would like to express my appreciation to those who assisted me in the preparation of this manual. In particular I would like to thank my secretary, Mrs. Claire Bartels, for her aid on many aspects of the preparation. Finally, I would like to thank Mario Triola for writing a fine statistics book.

CONTENTS

CHAPTER 2. DESCRIPTIVE STATISTICS

SOLUTIONS TO SECTION 2-2 SUMMARIZING DATA

2-1 The data is in whole integers, so the corresponding upper class limits would be 5, 8, 11, 14 and 17.

2-3 Since the data is presented with one decimal of accuracy, the associated upper class limits are 2.4, 4.9, 7.4, 9.9 and 12.4.

2-5 The differences between successive lower class limits is the class width. As a result, the successive lower class limits are 10, 30, 50, 70 and 90.

2-7 As in 2-5 above, we reach successive lower class limits by adding multiples of the class width to the lower limit of the first class. The lower class limits then are 7.5, 10, 12.5, 15, 17.5, 20, 22.5 and 25. With eight classes there are eight lower class limits.

2-9 The range of the data is 99 - 50 = 49. We divide the number of classes into the range and get 4.9. Rounding this up gives a class width of 5.

2-11 As in 2-9 dividing the range of 3.24 - .67 = 2.57 by the desired number of classes, 11, we get .2336. A reasonable number to round this up to would be .25.

2-13 One way to determine class width is that it is the difference between any two successive lower class limits. In this case that would be 88 - 80 = 8.

2-15 Subtracting successive lower class limits gives a class width of 21.2 - 16.2 = 5.

2-17 The class marks are the midpoint between the various upper and lower class limits. The marks are listed as:

IQ	Class Mark
80-87	83.5
88-95	91.5
96-103	99.5
104-111	107.5
112-119	115.5

2-19 As in 2-17, we obtain the class marks as midpoints between lower and upper class limits. They are:

Weight (kg)	Class Marks
16.2-21.1	18.65
21.2-26.1	23.65
26.2-31.1	28.65
31-2-36.1	33.65
36.2-41.1	38.65

2-21 We extend here on either side of each of the class limits by .5. Then our table of class boundaries will be:

IQ	Class Boundaries
80-87	79.5- 87.5
88-95	87.5- 95.5
96-103	95.5-103.5
104-111	103.5-111.5
112-119	111.5-119.5

2-23 We extend on either side of the two class limits by .05. The table of boundaries will be:

Weight (kg)	Class Boundaries
16.2-21.1	16.15-21.15
21.2-26.1	21.15-26.15
26.2-31.1	26.15-31.15
31.2-36.1	31.15-36.15
36.2-41.1	36.15-41.15

2-25 The cumulative frequency table is obtained by accumulating frequencies from class to class. The table would be given by:

IQ	Cumulative Frequency
Less than 88	16
Less than 96	53
Less than 104	103
Less than 112	132
Less than 120	146

2-27 As in 2-25 accumulating frequencies will give the following table:

Weight (kg)	Cumulative Frequency
Less than 21.2	16
Less than 26.2	31
Less than 31.2	43
Less than 36.2	51
Less than 41.2	54

2-29 Moving class width from 8 to 5 provides convenient classes. Using this provides a series of class limits given by:

IQ	
80-84	100-104
85-89	105-109
90-94	110-114
95-99	115-119

2-31 If we have the class 4, we would have the class limits as summarized below:

Weight (kg)
16-19.9
20-23.9
24-27.9
28-31.9
32-35.9
36-39.9
40-43.9

2-33 To find width of the classes we divide the range by the number of classes. This is (309 - 110)/10 = 19.9. We round this to 20. Lower limits of the first and second class would then be 110 and 130. Then lower and upper limits of the first class would be 110 and 129.

2-35 As in 2-33 we divide the range by the number of classes and get (42.7 - 17.3)/12 = 2.11. We would round this up to 2.2. Then successive lower class limits are 17.3 and 19.5. Then the lower and upper limits of the first class are 17.3 and 19.4 respectively.

2-37 The longest distance in the table is 40 hours, and the shortest is 1. Therefore, to compute class width we form

$$\frac{40 - 1}{8} = \frac{39}{8} = 4.75.$$

which we round up to 5. The table of classes with tally marks and frequencies is as follows:

Distances	Tally Marks	Frequencies
1-5	~~1111~~ ~~1111~~ 1	11
6-10	~~1111~~ 1111	9
11-15	~~1111~~ 1	6
16-20	~~1111~~	5
21-25	~~1111~~ 1	6
26-30		0
31-35	11	2
36-40	1	1

2-39 The highest consumption is 908 kilowatt hours while the lightest is 714 kilowatt hours. We compute the class width by

$$\frac{908 - 714}{10} = \frac{194}{10} = 19.4$$

which is rounded to 20. The table of classes, tally marks and frequencies is as follows:

Consumption	Tally Marks	Frequencies
710-729	111	3
730-749	1	1
750-769	1111	4
770-789	1111	4
790-809	11	2
810-829	111	3
830-849	~~1111~~ 11	7
850-869	1111	4
870-889	11	2
890-909	1	1

SOLUTIONS TO 2-2 B EXERCISES

2-41 Yes, there are some differences but such differences can rarely be detected using raw data only.

2-43 In this table the classes are of different widths.

SOLUTIONS FOR SECTION 2-3 - PICTURES OF DATA

2-45 a) For a histogram the horizontal scale consists of class boundaries. They are as given in the following table:

Score	Frequency	Boundaries
1-8	15	.5-8.5
9-16	12	8.5-16.5
17-24	7	16.5-24.5
25-32	5	24.5-32.5
33-40	1	32.5-40.5

As a result, the horizontal axis will consist of the numbers .5, 8.5, 16.5, 24.5, 32.5, and 40.5.

b) Class marks are used for the frequency polygon. As a result, the axis will consist of the numbers 4.5, 12.5, 20.5, 28.5, and 36.5.

c) For the ogive the horizontal axis numbers will be the same as in a) above.

2-47 The horizontal axis will consist of the class marks as given in b) above. The vertical axis must attain a frequency of 15. The frequency polygon is given by the following graph:

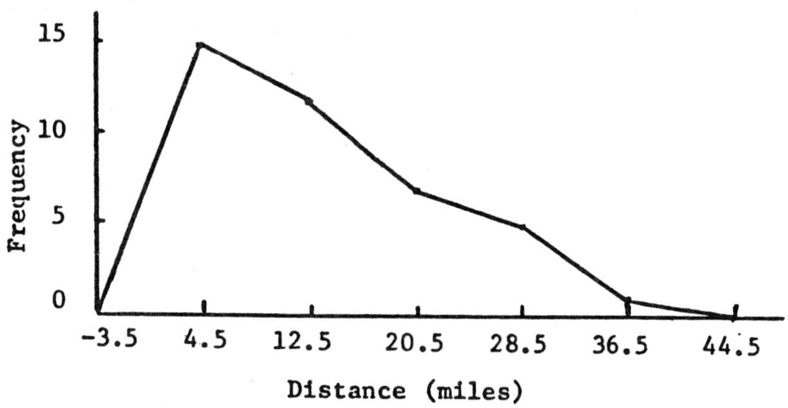

2-49 a) As in 2-45 above, we need the class boundaries. They are given in the following table.

Class Limits	Class Boundaries
0-3999	-.5-3999.5
4000-7999	3999.5-7999.5
8000-11999	7999.5-11999.5
12000-15999	11999.5-15999.5
16000-19999	15999.5-19999.5
20000-23999	19999.5-23999.5
24000-27999	23999.5-27999.5

b) The frequency polygon requires the class marks. The respective class marks are 1999.5, 5999.5, 9999.5, 13999.5, 17999.5, 21999.5, 25999.5.

c) For an ogive we would list the class boundaries as in a) above.

2-51 The horizontal axis is given in 2-49 b). We will need to have a vertical height of 19 since that is the maximum frequency. The graph is given below:

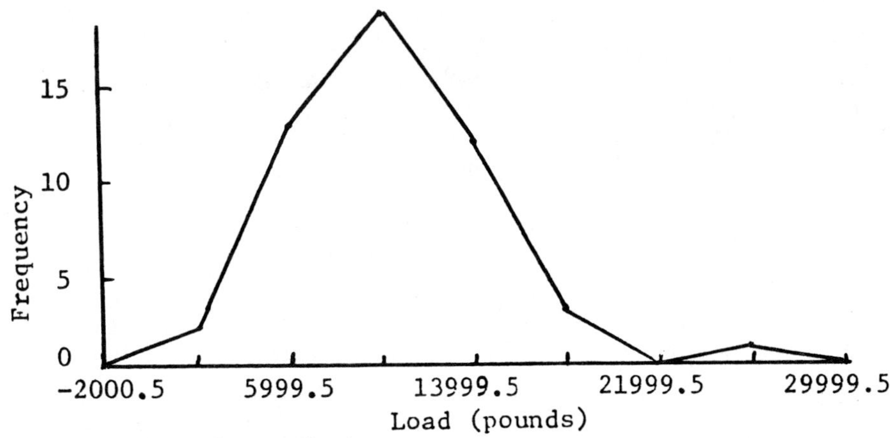

2-53 See 2-45 and 2-49 above for the method.

2-55 See 2-47 and 2-51 above for the method.

2-57 See 2-45 and 2-49 above for the method.

2-59 See 2-47 and 2-51 above for the method.

2-61 The first digit of each number in the data set is a 2. The other digits are in the list under leaves. The data set would then be 20, 20, 23, 25, and 28.

2-63 The first two digits of the numbers in the data set are in the stem column and the third is in the leaves column. The data set will be 406, 406, 407, 408, 410, 419, 419, 419, 419, 421, 423, 424, 426, 426, 430, 438, 438.

2-65 We shall use the tens digit as the stem and the units digit as leaves. The plot would then be:

Stem	Leaves
6	48
7	5679
8	000112334444569
9	0012233447

2-67 Using the first decimal digit in the stem and the second decimal digit as leaves we have:

Stem	Leaves
.5	000008
.6	55588
.7	055589
.8	0
.9	12

2-69 In the first row data goes from 0 to 9 dollars. In the second it proceeds from 10 to 19. This suggests use of class limits and associated frequencies as in the following table:

Dollar Sales	Frequency
0-9	28
10-19	22
20-29	20
30-39	22
40-49	21
50-59	11
60-69	8
70-79	0
80-89	4
90-99	1
100-109	2

We obtained the frequencies by counting the digit entries in the leaves. We used these digits as tally marks.

2-71 a) We read the graph to obtain frequencies.

Scores	Frequencies
15	1
16	3
17	2
18	7
19	6
20	9
21	8
22	3
23	5
24	2
25	1

b) The ogive uses the class boundaries along the horizontal axis. It attains a cumulative frequency of 45 along the vertical axis.

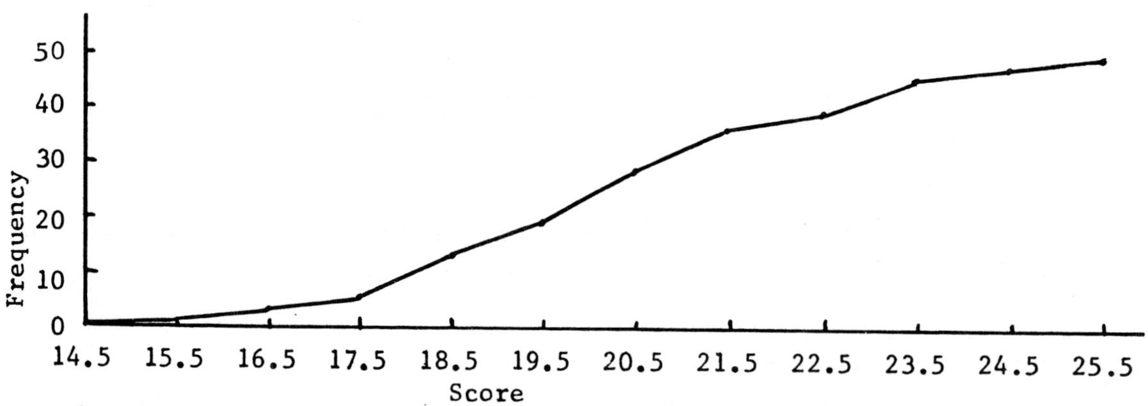

c) This would be the number scoring between 15 and 20. There are 28 of these.

d) 4 trainees failed out of the 47. This corresponds to

$$\frac{4}{47} = .085$$

or 8.5%.

SOLUTIONS FOR 2-3 B EXERCISES

2-73 a. The two histograms are as follows.

For π

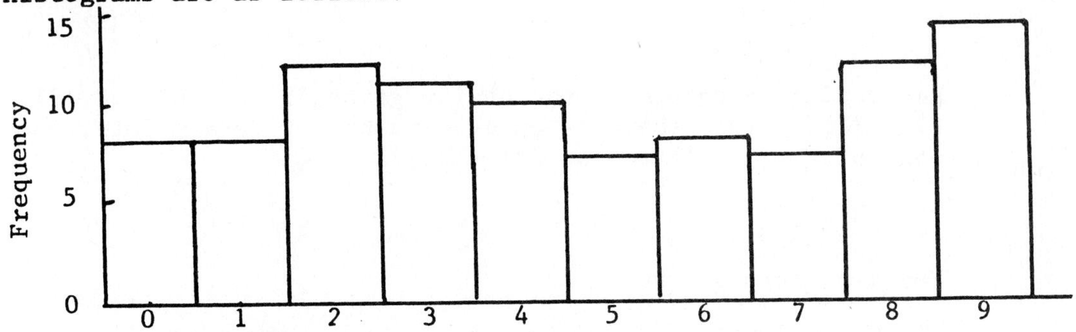

For 22/7

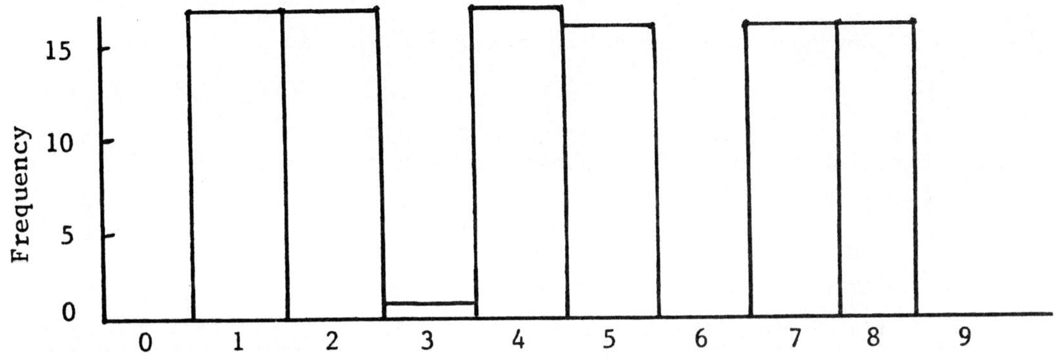

b. 22/7 has a repeating decimal expansion, but pi does not.

SOLUTIONS FOR SECTION 2-4 AVERAGES

2-77 To find the mean, we sum the numbers and divide by 20 since there are 20 data numbers. The mean

$$\bar{x} = \frac{3+3+3+3+5+2+3+3+3+2+4+2+2+3+2+3+5+3+4+4}{20}$$

$$= \frac{62}{20} = 3.1.$$

To get the median we must rearrange the data into increasing order of magnitude: 2,2,2,2,2,3,3,3,3,3,3,3,3,3,3,4,4,4,5,5.

The median is the point midway between 3 and 3 since there are an even number of data points. The median is 3.

The mode is the most frequently occurring data point and is 3.

The midrange is

$$\frac{2+5}{2} = \frac{7}{2} = 3\frac{1}{2}.$$

2-79 The mean is given by

$$\bar{x} = \frac{74 + 73 + 77 + 77 + 71 + 68 + 65 + 77 + 67 + 66}{10}$$

$$= \frac{715}{10} = 71.5.$$

Reordering to determine the median gives 65, 66, 67, 68, 71, 73, 74, 77, 77, 77. Since there is an even number of data points, the median is the mean of the middle two or (71 + 73)/2 = 72.

The mode is 77 since there are more of these than any other number.

The mid midrange is

$$\frac{65 + 77}{2} = \frac{142}{2} = 71 .$$

2-81 Proceed as in 2-77 to obtain a mean of 9.3 , a mode of 9, a median of 9, and a midrange of 11.

2-83 The mean is

$$\bar{x} = \frac{8.8+8.7+6.6+11.3+8.4+8.1+6.5+5.2+8.4+6.3+7+11.6+9+6.2+10.3+10.9}{16}$$

$$= \frac{133.3}{16} = 8.33.$$

We reorder to obtain the median: 5.2, 6.2, 6.3, 6.5, 6.6, 7.0, 8.1, 8.4, 8.4, 8.7, 8.8, 9.0, 10.3, 10.9, 11.3, 11.6. Because there are an even number of data points the median is the mean of the middle two points. That is, it is the mean of 8.4 and 8.4, or 8.4.

The mode is 8.4 and the midrange is

$$\frac{5.2 + 11.6}{2} = \frac{16.8}{2} = 8.4.$$

2-85 Similar in method to 2-83.

2-87 The mean is

$$\bar{x} = \frac{.74+.71+.41+.82+.74+.85+.99+.71+.57+.85+.57+.55}{12}$$

$$= \frac{8.51}{12} = .709.$$

When the data is reordered to determine the median, the resultant data list is: .41, .55, .57, .57, .71, .71, .74, .74, .82, .85, .85, .99. There are an even number of these so the median is the mean of the middle two. This is (.71 + .74)/2 = .725. In this data set, there are four values repeated twice. As a result, each is considered a mode. The modes are then .57, .71, .74, and .85. The midrange is (.41 + .99)/2 = .70.

2-89 The mean is

$$\bar{x} = \frac{\text{The sum of all of the distances}}{\text{The number of distances}}$$

$$= \frac{533}{40} = 13.325.$$

Reordering the data in increasing order gives: 1,1,1,1,2,3,4,5,5, 5,5,6,8,8,8,9,9,10,10,10,12,12,14,15,15,15,16,18,20,20,20,21,23,24, 25,25,25,31,31,40. The median is the midpoint between 10 and 12. It is (10 + 12)/2 = 11. The modes are 1 and 5 since both occur 4 times, and the midrange is (1 + 40)/2 = 20.5.

2-91 This exercise is solved in the same manner as 2-89.

2-93 a) The table of frequencies and class marks is as follows.

Time	Class Mark X	Frequency f	x•f
0- 39	19.5	17	331.5
40- 79	59.5	41	2439.5
80-119	99.5	80	7960.
120-159	139.5	49	6835.5
160-199	179.5	4	718.
Totals		191	18284.5

b) The mean is the sum of column x • f divided by the sum of the frequencies. Thus $\bar{X}$ = 18284.5/191 = 95.73.

2-95 a) The table of frequencies and marks is as follows. In this case marks and family sizes are the same.

Family Size	Frequency f	x•f
2	51	102
3	31	93
4	27	108
5	12	60
6	4	24
7	1	7
8	1	8
Totals	127	402

b) The mean will be $\bar{X}$ = 402/127 = 3.17.

2-97 a) A table of class marks, frequencies and their products is as follows.

Index	Mark X	Frequency f	x•f
.10-.11	.105	9	.945
.12-.13	.125	16	2.0
.14-.15	.145	12	1.74
.16-.17	.165	3	.495
.18-.19	.185	1	.185
Totals		41	5.365

b) The mean would then be 5.365/41 = .131.

2-99 This exercise is solved in a manner similar to 2-97.

SOLUTIONS TO 2-4 B EXERCISES

2-101 a. The mean is

$$\bar{x} = \frac{1023.6 + 1024.2 + 1026.7 + 1033.5 + 1040.3}{5} = 1029.66$$

The median is 1026.7. There is no mode since no reading appears more than once. The midrange is (1023.6 + 1040.3)/2 = 1031.95.

b. The revised data is: 23.6, 24.2, 26.7, 33.5, 40.3. The new mean is

$$\bar{X} = \frac{23.6 + 24.2 + 26.7 + 33.5 + 40.3}{5}$$

$$= 29.66.$$

The median is 26.7 There is no mode, and the midrange is (23.6 + 40.3)/2 = 31.95.

c. Yes, each differs by the factor of 1000.

d. They will differ by the corresponding constant k.

e. Using the computations as in a. and b., the resulting mean will be .02966, the median is .0267, there is no mode, and the midrange is .03195.

f. The resulting scores will be 236, 242, 267, 335, 403. The mean will be 296.6, the median will be 267, there is no mode, and the midrange is 319.5.

g. Yes, they differ by a multiple of 10000.

h. They will differ by the corresponding multiple k.

i. Computations can sometimes be simplified by adding, subtracting, multiplying, or dividing the data by a constant.

2-103 No, the means are the same but the two data sets are clearly different. The data from 2-80 is more variable or spread out.

2-105 The mean is calculated in a manner similar to 2-97. Usually classes with larger class intervals will give poorer estimates of means.

2-107 The product of the numbers is 1.5518 . Since there are 10 numbers, the geometric mean is

$$\sqrt[5]{1.5518} = 1.092.$$

2-111 Referring to the data in table 2-3, there are a total of n = 150 homes. The median class would be the one from 125000 - 149999. In that case the frequency of the median class is 47, while the sum of the frequencies of the classes preceeding the median class is 32. The lower limit of the median class is 125000. Then the median is

$$125000 - (25000)\left[\frac{75.5 - (33)}{47}\right] = 147606.$$

SOLUTIONS FOR SECTION 2-5 - DISPERSION STATISTICS

2-113 The range is the largest minus the smallest which is 5 - 2 = 3. The sum of the observations is 62. We then square the observations and get 9,9,9,9,25,4,9,9,9,4,16,4,4,8,4,9,25,9,16,16. The sum of these squares is 208. By using formula 2-5 the variance is:

$$s^2 = \frac{n(\sum X^2) - (\sum X)^2}{n(n-1)}$$

$$= \frac{20(208) - (62)^2}{20(19)}$$

$$= \frac{4160 - 3844}{380} = \frac{316}{380} = .831.$$

The standard deviation is $s = \sqrt{.779} = .883$.

2-115 The range is 77 - 65 = 12. To compute the variance we need $\sum X = 715$. In addition we need $\sum X^2 = 51327$. Then

$$s^2 = \frac{10(51327) - (715)^2}{10(9)} = 22.7.$$

Also, the standard deviation is $s = \sqrt{22.7} = 4.77$.

2-117 The range is 15 - 7 = 8. We will use formula 2-5. We have the sum of the days is $\sum X = 93$. We get the sum of squares as $\sum X^2 = 100 + 81 + 64 + 64 + 81 + 81 + 49 + 225 + 81 + 81 = 907$. Then using formula 2-5, the variance is

$$s^2 = \frac{n(\sum X^2) - (\sum X)^2}{n(n-1)}$$

$$= \frac{10(907) - (93)^2}{90} = 4.68.$$

Then the standard deviation is $s = \sqrt{4.68} = 2.16$.

2-119 As above, the range is the largest minus the smallest, which in this case

11.6 - 5.2 = 6.4. For the variance we need $\sum X = 133.3$. We accumulate the squares of the observations and get $\sum X^2 = 1167.99$. Then the variance is given by

$$s^2 = \frac{n(\sum X^2) - (\sum X)^2}{n(n-1)}$$

$$= \frac{16(1167.99) - (133.3)^2}{16(15)} = 3.83.$$

The standard deviation is $s = \sqrt{3.83} = 1.96$.

2-121 The range is 1.824 - 1.013 = .811. For the variance we need $\sum X = 12.005$. The sum of squares of the observations are given by $\sum X^2 = 16.665$. Then the variance is given by

$$s^2 = \frac{n(\sum X^2) - (\sum X)^2}{n(n-1)} = \frac{9(16.665) - (12.005)^2}{9(8)} = .081.$$

The standard deviation is $s = \sqrt{.081} = .285$.

2-123 The range is .99 - .41 = .58. To obtain the variance we need $\sum X = 8.51$. In addition $\sum X^2 = 6.3213$. Then the variance is

$$s^2 = \frac{12(6.3213) - (8.51)^2}{12(11)} = .026.$$

The standard deviation is $s = \sqrt{.026} = .161$.

2-125 The range is 40 - 1 = 39 miles. To find the variance and standard deviation using formula 2-5 we need $\sum X$ and $\sum X^2$. We have $\sum X = 533$. In addition $\sum X^2 = 10705$. Then we compute the variance by

$$s^2 = \frac{n(\sum X^2) - (\sum X)^2}{n(n-1)} = \frac{40(10705) - (533)^2}{(40)(39)} = 92.38.$$

The standard deviation is $s = \sqrt{92.38} = 9.61$.

2-127 This exercise is solved in a manner similar to 2-125.

2-129 The range, variance and standard deviation will all remain the same.

2-131 Multiplying each observation by 10 will yield data of 100, 90, 80, 80, 90, 90, 70, 150, 90, 90. The range is 150 - 70 = 80 which is 10 times the original range. The value of $\sum X = 930$ which is 10 times the original value and $\sum X^2 = 90700$ which is 100 times the original value. Then

$$s^2 = \frac{10(90700) - (930)^2}{(10)(9)} = 468.$$

The standard deviation is $s = 21.6$. Then s^2 is 100 times and s is 10 times the respective original values.

2-133 To find the variance and standard deviation, we will use a table to aid the computations.

	Class Marks X	Frequency f	xf	x^2	x^2f
	19.5	17	331.5	380.25	6464.25
	59.5	41	2439.5	3540.25	145150.25
	99.5	80	7960.	9900.25	792020.
	139.5	49	6835.5	19460.25	953552.25
	179.5	4	718.	32220.25	128881.
Totals		191	18284.5		2026067.8

The variance is given by

$$s^2 = \frac{n(\sum x^2f) - (\sum xf)^2}{n(n-1)}$$

$$= \frac{191(2026067.8) - (18284.5)^2}{191(190)}$$

$$= 1451 \quad \text{and } s = \sqrt{1451} = 38.1.$$

2-135 As above, we use a table to assist our computations.

Class Marks x	Frequency f	xf	x^2	x^2f
2	51	102	4	204
3	31	93	9	279
4	27	108	16	432
5	12	60	25	300
6	4	24	36	144
7	1	7	49	49
8	1	8	64	64
Totals	127	402		1472

Then the variance is

$$s^2 = \frac{n(\sum x^2f)-(\sum xf)^2}{n(n-1)} = \frac{127(1472) - 402^2}{127(126)} = 1.58$$

The standard deviation is $s = \sqrt{1.58} = 1.26$.

2-137 Using our table we have

Class Marks x	Frequency f	xf	x^2	x^2f
.105	9	.945	.011	.099
.125	16	2.0	.016	.256
.145	12	1.74	.021	.252
.165	3	.495	.027	.081
.185	1	.185	.034	.034
Totals	41	5.365		.722

Then $s^2 = \dfrac{41(.722) - (5.365)^2}{41(40)} = .0005.$

and $s = \sqrt{.0005} = .022.$

2-139 This exercise is solved in the same manner as 2-137.

2-141 The group of statistics students as college students would normally be a more homogeneous group. Hence, their I.Q. scores should have a smaller standard deviation.

2-143 If all scores are the same, the standard deviation will be zero. Standard deviations are sums of squares which could all be zero but never negative.

SOLUTIONS TO 2-5 B EXERCISES

2-145 The range for group C is 20-1 = 19 while that of group D is 18-2 = 16. As a result group C has the greater range. For group C the variance is

$$s_C^2 = \frac{n(\sum x^2) - (\sum x)^2}{n(n-1)} = \frac{10(3601) - (181)^2}{10(9)} = 36.1$$

and for group D is

$$s_D^2 = \frac{10(1380) - (100)^2}{10(9)} = 42.2.$$

The two standard deviations are $s_C = 6.01$ and $s_D = 6.5$.

The variance and standard deviation of the C group is less. Since the overall variability of the C group appears less, it would better in this case to use the variance or standard deviation than the range.

2-147 a. The interval from 52 to 60 is the mean minus one standard deviation to the mean plus one standard deviation. Since this is the case, the interval encompasses about 68% of the data.

b. The interval 8 units or less units from the mean will be the data within 2 standard deviations from the mean. Therefore, this would encompass 95% of the data.

c. This would be the data within 3 standard deviations of the mean. That is it would be within 12 units. This would be the interval from 44. to 68.

2-151 Since a shift will not affect the standard deviation, we can subtract 17 from each data point and have the same value of σ. The data would become 1, 2, 3, ... , 165. Then

$$\sigma = \sqrt{\frac{n^2-1}{12}} = \sqrt{\frac{165^2-1}{12}} = 47.63.$$

The mean of the data 1, 2, 3, ..., 165 would be

$$\mu = \frac{1+2+3+...+n}{n} = \frac{1+2+3+...+165}{165} = \frac{\frac{n(n+1)}{2}}{n} = \frac{\frac{165(166)}{2}}{165}$$

$$= \frac{166}{2} = 83.$$

This would be 17 less than the mean of the original data. Therefore, $\mu = 100$.

2-153 Upon looking back at those problems where the data was transformed, we find the following result.

1. The mean would become $\bar{x}_C = \frac{5}{9}(\bar{x}_F - 32) = \frac{5}{9}(40.2-32) = 4.56.$

2. The variance would become $s_C^2 = (5/9)^2 s_F^2$. A shift of the data does not affect its variability. Then $s_C^2 = 2.78$.

3. The standard deviation will be $s_C = (5/9)s_F = 1.67$.

2-155 a. The coefficient of variation is

$$\frac{\sigma}{\mu} 100 = \frac{15}{100} 100 = 15.$$

b. This is $\frac{50}{250} \bullet 100 = 20.$

c. We compute the sample mean as 105/10 = 10.5. The sample standard deviation is 10. Then the coefficient of variation is 100(10/10.5) = 95.2.

SOLUTIONS FOR SECTION 2-6 - MEASURES OF POSITION

2-157 a. The z score is

$$z = \frac{x - \bar{x}}{s} = \frac{80 - 60}{10} = \frac{20}{10} = 2.$$

b. Here

$$z = \frac{x - \bar{x}}{s} = \frac{40 - 60}{10} = \frac{-20}{10} = -2.$$

c. The z score here is

$$z = \frac{x - \bar{x}}{s} = \frac{65 - 60}{10} = \frac{5}{10} = .5.$$

2-159 The z score using the formula is

$$z = \frac{x - \bar{x}}{s} = \frac{10. - 7.06}{5.32} = .553.$$

2-161 For the z score we have

$$z = \frac{2.00 - 2.97}{.60} = \frac{-.97}{.60} = -1.62.$$

2-163 The z score is as follows.

$$z = \frac{10. - 22.83}{8.55} = \frac{-12.83}{8.55} = -1.50.$$

2-165 The z score of part a) is

$$z = \frac{53 - 50}{10} = \frac{3}{10} = .3.$$

The z score of part b) is

$$z = \frac{53 - 50}{5} = \frac{3}{5} = .6.$$

A z score of .6 would have the higher relative position.

2-167 The three respective z scores are the means for comparison. They are

a) $z = \frac{60 - 50}{5} = \frac{10}{5} = 2.$

b) $z = \frac{230 - 200}{10} = \frac{30}{10} = 3.$

c) $z = \frac{540 - 500}{15} = \frac{40}{15} = 2.67.$

The z score of b) is the best of the three.

2-169 Of the 150 scores, 14 are less than 110000. As a result, the percentile is 14/150 x 100 = 9.3 or approximately 9 th percentile.

2-171 There are 110 scores less than 175000 so the corresponding percentile is 110/150 x 100 = 73.3 or approximately the 73 rd percentile.

2-173 Following the flowchart, we compute

$$L = \frac{(15) \times (150)}{100} = 22.5.$$

We find the 23rd score which is 117500 which is then the 15th percentile.

2-175 Using the steps of the flowchart of Figure 2-10 we have

$$L = \frac{(80) \times (150)}{100} = 120.$$

Since this is an integer, then the 80th percentile will be the mean of the 120th and the 121st prices. These are 183000 and 185000. Then the mean is 184000.

2-177 Since Q_1 is the same as P_{25}, we will use our procedure to find that score. We have

$$L = \frac{(25)(150)}{100} = 37.5.$$

Then P_{25} will be the 37th score because L was not an integer. That will be 126500.

2-179 The third quartile Q_3 is the same as P_{75}. To find that we set

$$L = \frac{(75)(150)}{100} = 112.5.$$

Since L is not an integer, D_3 will be the 113th score. That is Q_3 = 179000.

2-181 There are 4 of the 50 weights less than 5030 lb. Therefore, the percentile is (4/50)100 = 8.

2-183 There are 36 weights less than 12430. Therefore, the percentile is (36/50)100 = 72.

2-185 We use the flowchart and compute L as

$$L = \frac{(15)(50)}{100} = 7.5.$$

Since L is not an integer, we move up to the nearest integer which is 8. Then the 8th lowest weight is 6400.

2-187 This exercise is solved like 2-185.

2-189 The first quartile is the 25 th percentile. We compute L as

$$L = \frac{(25)(50)}{100} = 12.5.$$

Since this is not an integer, we need only find the 13th lowest weight which is 7470.

2-191 The third decile is the 30th percentile. The 30th percentile is computed using the same method as 2-189.

SOLUTIONS TO 2-6 B EXERCISES

2-193 a. The first quartile was found in 2-177 to be 126500 and the third quartile from 2-179 was 179000. So the interquartile range is 179000 - 125600 = 52500.

b. The second quartile is 152750.

c. The difference between P_{10} and P_{90} is 96703.

d. By definition both answers are yes.

e. No.

SOLUTIONS TO THE REVIEW EXERCISES

2-199 a) The mean is

$$\bar{x} = \frac{\sum x}{n} = \frac{19+18+12+25+22+8+8+16}{8} = 16.$$

b. We place data in increasing order 8,8,12,16,18,19,22,25. Since there is an even number of data points, the median would be (16+18)/2 = 17.

c. There are two 8 s so the mode is 8.

d. The midrange is (8+25)/2 = 16.5.

e. The range is 25-8 = 17.

f. The variance is

$$s^2 = \frac{n\sum x^2-(\sum x)^2}{n(n-1)} = \frac{8(2322)-(128)^2}{8(7)} = 39.1.$$

g. The standard deviation $s = \sqrt{39.1} = 6.26$.

2-201 We first determine class width. We form

$$\frac{\text{largest - smallest}}{10} = \frac{97 - 8}{10} = 8.9.$$

Rounding this up to 9, we have determined our class width. The sequence of lower class limits would be 8, 17, 26, ... The construction of the table would be given below.

Number Of Days	Tally Marks	Frequency
8-16	1111	4
17-25	~~1111~~ 1	6
26-34	~~1111~~ 111	8
35-43	1111	4
44-52	~~1111~~ 1	6
53-61	1111	4
62-70	1111	4
71-79	11	2
80-88	~~1111~~ 11	7
89-97	~~1111~~	5

2-203 a. To find Q_1 we determine that point below which 25% of the data lies. There are 50 data points. We form L = (25x50)/100 = 12.5. We round up to 13 and find the 13th lowest score which is 27.

b. Form L = (45x50)/100 = 22.5 which we round up to 23. We determine the 23rd lowest score which is 44.

c. We determine the percentile of 30 by noting that there are 16 lower scores. Consequently we have 30 as the (16/50)x100 = 32th percentile.

2-205 a. To find the mean we sum the sizes and divide by the number. That is $\bar{x} = (\sum x)/n = (22221)/12 = 1851.75$.

b. For the median, we put the data in increasing order of magnitude and find the mean of the middle two since there is an even number of data points. In order the data is 1200, 1248, 1300, 1632, 1600, 1800, 1956, 2000, 2000, 2025, 2400, 3060. Then the median is the mean of 1800 and 1956 which is 1878.

c. The mode is 2000 since there are two of those.

d. The midrange is the mean of 1200 and 3060 which is 2130.

e. The range is 3060 - 1200 = 1860.

f. We compute the variance using the formula

$$s^2 = \frac{n\sum x^2 - (\sum x)^2}{n(n-1)} = \frac{12(44201089) - (22221)^2}{12(11)}$$

$$= 277577.$$

g. The standard deviation is then

$$s = \sqrt{\text{variance}} = \sqrt{277577} = 526.9.$$

2-207 a. The z score is given by

$$z = \frac{x - \bar{x}}{s} = \frac{1200 - 1851.75}{526.9} = -1.24.$$

b. The z score is given by

$$z = \frac{x - x}{s} = \frac{2400 - 1851.75}{526.9} = 1.04.$$

2-209

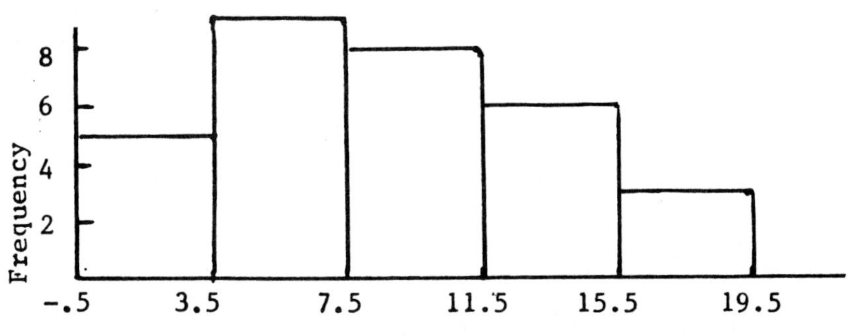

2-211

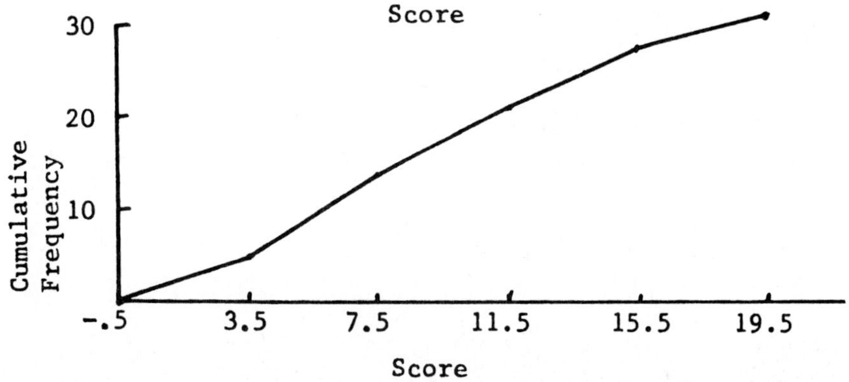

2-213 a. For the mean we sum the data and divide by the number of data values. That is $\bar{x} = (1972)/11 = 179.3$

b. We arrange the data in increasing order as 168, 170, 174, 175, 181, 181, 182, 183, 183, 184, 191. Then the median is the middle score which is

181.

c. There are two modes 181 and 183.

d. The midrange is the mean of 168 and 191. That is, the midrange is (168 + 191)/2 = 179.5.

e. The range is 191 - 168 = 23.

f. For the variance we use

$$s^2 = \frac{n\sum x^2 - (\sum x)^2}{n(n-1)} = \frac{11(353986) - (1972)^2}{11(10)}$$

$$= 46.$$

g. The standard deviation is the square root of the variance which is 6.78.

2-215 For the stem and leaf diagram we have

Stem	Leaf
16	8
17	0,4,5
18	1,1,2,3,3,4
19	1

CHAPTER 3. PROBABILITY

SOLUTIONS TO SECTION 3.2 - FUNDAMENTALS

3-1 Since probabilities must be numbers between 0 and 1 including both 0 and 1, the numbers which could not be probabilities in the given list are 1.2, 77/75, -1/2, 5, 1.001, and the square root of 2.

3-3 Since any student is as likely to be generated as any other, the probability if 1/30.

3-5 There were 5 answers possible with one correct therefore, the probability will be 1/5.

3-7 Since there were 400 drivers with 136 in an accident, the probability is 136/400 = .34.

3-9 This is an empirical probability which because of three failures in eight helicopters is 3/8.

3-11 Because there are 56 unlisted and 144 listed, there were 200 total telephone numbers. Then the probability that a randomly selected one of these is unlisted is 56/200 = 7/25.

3-13 There is a total of 1500 records. Since 135 were country music,the probability is 135/1500 = .09.

3-15 A total of 55 did not awake among the 95 subjects, the probability of selecting someone who did not awake is 55/95 = 11/19.

3-17 Since there were 44 million Republicans and 36 million Democrats, there were a total of 80 million voters. Among the 80 million voters, 44 million were Republican. Therefore, the probability is 44/80 = 11/20.

3-19 There were a total of 235 + 755 = 990 subjects. The characteristic was present in 235 subjects. Therefore, the probability is 235/990 = 47/198.

3-21 There were a total of 90 + 80 + 20 + 10 = 200 people sampled, and 10 had type AB blood. Therefore, the probability is 10/200 = 1/20.

3-23 There were a total of 10589 + 22551 + 636 + 963 = 34739 drivers. Of those 636 were DWI convictions in Dutchess County. As a result, the probability is 636/34739 = .0183.

3-25 a) We list these as ordered pairs with the first being the first child and the second as the second child. The possibilities are

boy-boy, boy-girl,
girl-boy, girl-girl.

b) Only one of the 4 equally likely outcomes consists of two boys. The probability would therefore be 1/4.

c) Two outcomes have exactly one child of each sex. Therefore, the probability is 2/4 or 1/2.

3-27 a) The list of possible solutions is:

T T T
T T F
T F T
F T T
F F T
F T F
T F F
F F F

b) Exactly one of these sequences contains all three answers correct. As a result, the probability is 1/8.

c) Exactly one of these sequences contains all three answers wrong. As a result, the probability is 1/8.

d) If at least two answers are correct, we must include all three correct as well as exactly two correct. There is one way for all three to be correct, and three ways to have exactly one wrong. Therefore there are four ways to get at least two right. As a result, the probability will be 4/8 or 1/2.

SOLUTIONS TO 3-2 B EXERCISES

3-29 a) The possible outcomes can be seen more clearly if we designate the parents as P1, P2, and P3 and the various genes as P1R for the red gene of parent 1. When a parent has 2 genes of the same color, we will write, for example, P2W1 as the first white gene of parent 2. The possible outcomes are then:

Specification	In Short Form
P1R-P2R-P3R1	R-R-R
P1R-P2R-P3R2	R-R-R
P1R-P2R-P3W	R-R-W
P1R-P2W1-P3R1	R-W-R
P1R-P2W1-P3R2	R-W-R
P1R-P2W1-P3W	R-W-W
P1R-P2W2-P3R1	R-W-R
P1R-P2W2-P3R2	R-W-R
P1R-P2W2-P3W	R-W-W
P1W-P2R-P3R1	W-R-R
P1W-P2R-P3R2	W-R-R
P1W-P2R-P3W	W-R-W
P1W-P2W1-P3R1	W-W-R
P1W-P2W1-P3R2	W-W-R
P1W-P2W1-P3W	W-W-W
P1W-P2W2-P3R1	W-W-R
P1W-P2W2-P3R2	W-W-R
P1W-P2W2-P3W	W-W-W
P1B-P2R-P3R1	B-R-R
P1B-P2R-P3R2	B-R-R

P1B-P2R-P3W B-R-W
P1B-P2W1-P3R1 B-W-R
P1B-P2W1-P3R2 B-W-R
P1B-P2W2-P3W B-W-W
P1B-P2W2-P3R1 B-W-R
P1B-P2W2-P3R2 B-W-R
P1B-P2W2-P3W B-W-W

b) There are two outcomes with blue-red-red, so the probability is 2/27.

c) Counting the ways to get at least two white genes among parents, there are 11 of these. As a result, the probability is 11/27.

d) No children have a majority of blue genes. There are 5 cases where all three genes are present. The probability is 5/27.

e) By part c), the probability of white is 11/27. The probability of not white added to the probability of white is 1, since the antenna is either white or it is not white. As a result, the probability of not white is 1 - (the probability of white). That is, the probability of not white is 1 - (11/27) = 16/27.

3-31 The probability that they lie on the same hemisphere is the same as the probability that the central angle between the two is less than or equal to 180°. That follows because placing one fly in a fixed location and considering all hemispheres containing that fly (not necessarily centered at the fly), we see that every point in the sphere is in <u>some</u> hemisphere containing that fly so the probability is 1.0.

SOLUTIONS FOR SECTION 3.3 - ADDITION RULE

3-33 Only in b are the events exclusive.

3-35 If l is the event of less than 100 and G is the event of greater than 900, then by the addition formula

$$P(l \text{ or } G) = P(l) + P(G) - P(l \text{ and } G)$$
$$= \frac{100}{1000} + \frac{99}{1000} - 0 = \frac{199}{1000}$$

since there are 100 numbers from 0 to 99 and 99 numbers between 901 and 999.

3-37 If N is the event of a november birth date and D is the event of a december one, then

$$P(N \text{ or } D) = P(N) + P(D) - P(N \text{ and } D)$$
$$= \frac{30}{366} + \frac{31}{366} = \frac{61}{366}.$$

The two events are mutually exclusive so P(N and D) = 0.

3-39 There are a total of 43 companies. If M is the event of selecting a mining company and C of selecting a construction company, then

$$P(M \text{ or } C) = P(M) + P(C) - P(M \text{ and } C).$$

$$P(M) = \frac{12}{43},\quad P(C) = \frac{18}{43},\quad P(M \text{ and } C) = 0.$$

The latter is true, because a randomly selected company could not be both a mining and a construction company. Thus,

$$P(M \text{ or } C) = \frac{12}{43} + \frac{18}{43} = \frac{30}{43}.$$

3-41 Let B_1 be the event of selecting a resident under 10 years old, and B_2 the event of selecting someone over 69. These two events are mutually exclusive because people cannot be both ages at the same time. By the addition formula

$$P(B_1 \text{ or } B_2) = P(B_1) + P(B_2)$$

$$= \frac{42284}{222295} + \frac{14625}{222295} = \frac{56909}{222295} = .256.$$

3-43 If A is the event of selecting someone under 30 and B of selecting someone between 20 and 49, then by the addition rule we have

$$P(A \text{ or } B) = P(A) + P(B) - P(A \text{ and } B)$$

where (A and B) is the event of selecting someone under 30 and between 20 and 49. That is it is someone between 20 and 29. Then,

$$P(A) = \frac{114476}{222295},\quad P(B) = \frac{86740}{222295},\quad P(A \text{ and } B) = \frac{32010}{222295}.$$

Thus

$$P(A \text{ or } B) = \frac{114476}{222295} + \frac{86740}{222295} - \frac{32010}{222295} = \frac{169206}{222295} = .761.$$

3-45 By the addition rule

P(Woman) = P(Woman with a speeding ticket) + P(Woman with parking) - P(Woman with both speeding and parking ticket)

P(Woman with a speeding ticket) = 50/150.

P(Woman with a parking ticket) = 50/140.

No one received both types of tickets.
P(Woman) = 50/150 + 50/140 - 0 = 10/29.

3-47 Let M be the event that the selected person is a man and S that the person received a speeding ticket. Since there were 290 people,

P(M) = 190/290 while P(S) = 150/290.

P(M and S) = 100/290 since there are 100 men who received speeding tickets. Therefore,

$$P(M \text{ or } S) = P(M) + P(S) - P(M \text{ and } S)$$
$$= 190/290 + 150/290 - 100/290 = 240/290 = 24/29.$$

3-49 This exercise is solved in the same manner as 3-45.

3-51 This exercise is solved in the same manner as 3-47.

3-53 By the addition rule,

P(group A or Group B) = P(group A) + P(Group B)

$$= \frac{40}{100} + \frac{10}{100} = \frac{50}{100} = .5.$$

3-55 By the addition formula,

P(Group A or Rh-) = P(group A) + P(Rh-) - P(Group A and Rh-)

$$= \frac{40}{100} + \frac{14}{100} - \frac{5}{100} = \frac{49}{100} = .49.$$

3-57 Let A be the event the person is in Group A, B that they are in Group B and C that they are in Group AB. Then since these events are independent,

$$P(A \text{ or } B \text{ or } C) = P(A) + P(B) + P(C)$$

$$= \frac{40}{100} + \frac{10}{100} + \frac{5}{100} = \frac{55}{100} = .55$$

3-59 This exercise is solved as in 3-55.

SOLUTIONS TO 3-3 B EXERCISES

3-61 Using the addition formula

$$P(A \text{ or } B) = P(A) + P(B) - P(A \text{ and } B)$$

$$\frac{1}{3} = P(A) + \frac{1}{4} - \frac{1}{5}$$

As a result,

$$P(A) = \frac{1}{3} + \frac{1}{5} - \frac{1}{4} = \frac{20}{60} + \frac{12}{60} - \frac{15}{60} = \frac{17}{60}.$$

3-63 By the addition formula

$$P(A \text{ or } B) = P(A) + P(B) - P(A \text{ and } B)$$

or

$$\frac{79}{120} = \frac{11}{24} + \frac{13}{40} - P(A \text{ and } B)$$

3-65 a. Since A and B are mutually exclusive, then P(A and B) = 0. Therefore, P(A or B) = P(A) + P(B) = .9.

b. We have P(A or B) = P(A) + P(B) - P(a and B)
= .4 + .5 - P(A and B).
Since A and B are not mutually exclusive, P(A and B) $\neq$ 0, then P(A or B) < .9.

c. By the addition formula

$$P(A \text{ or } B) = P(A) + P(B) - P(A \text{ and } B)$$

or

$$.8 = .4 + P(B) - P(A \text{ and } B).$$

Consequently, we can conclude that P(B) is at least .4. Since P(A and B) $\leq$ P(A) = .4, then P(B) is at most .8.

3-67 Using a Venn diagram, the exclusive or excludes the event (A and B). Since event A includes (A and B) as does event B, then the formula to totally exclude (A and B) is

$$P(A \text{ or } B) = P(A) + P(B) - 2P(A \text{ and } B).$$

SOLUTIONS FOR SECTION 3.4 - MULTIPLICATION RULE

3-71 Let A be the event that the first person is married and B that the second one is married. Due to independence of these two events,

$$P(A \text{ and } B) = P(A)P(B) = (.62)(.62) = .3844.$$

3-73 Let A, B, C, and D be the events of the individual injuries involve arms and hands. Since these events are independent, the probability is given by

$$P(A \text{ and } B \text{ and } C \text{ and } D) = P(A)P(B)P(C)P(D)$$
$$= (.57)^4 = .10556.$$

3-75 Assuming independence of the 3 separate events,

$$P(\text{each one swims}) = (.18)^3 = .0058.$$

3-77 Let A be the event that the first chip is defective and B that the second is defective. Then the probability that all both are defective is

$$P(A \text{ and } B) = P(A)P(B|A)$$
$$= 2/6 \bullet 1/5 = 2/30 = 1/15.$$

3-79 Assuming independence of the events of the three selected people, then each has a probability of 1/365 of being born on the fourth of July. Then,
$$P(A \text{ and } B \text{ and } C) = P(A)P(B)P(C) = (1/365)^3.$$

3-81 For A, B, C, and D, the events of selecting months by the four firms, then the probability that A and B are different months is

$$P(A \text{ and } B) = P(A)P(B \text{ is some month not the same as } A)$$

3-81 For A, B, C, and D, the events of selecting months by the four firms, then the probability that A and B are different months is

P(A and B) = P(A)P(B is some month not the same as A)

= 12/12 • 11/12 = 11/12.

The 11/12 probability follows because firm A has selected a month. Following this line of reasoning,

P(all 4 are different months) = 12/12 • 11/12 • 10/12 • 9/12

= 165/288 = 55/96.

3-83 a) If A is the event that the first digit is odd, and B that the second is, then

P(A and B) = P(A)P(B|A) = 3/5 • 2/4 • = 6/20 = 3/10 = .3.

P(B|A) = 2/4 since B|A means A is given as some specific number. Thus there are only two odd numbers and four digits left after A has been determined.

b) In this part the value of A has no bearing on possible values of B, and hence events A and B are independent. Therefore,

P(A and B) = P(A)P(B) = 3/5 • 3/5 = 9/25 = .36.

3-85 If we assume independence of the selections, then
P(each of the 10 thermometers is in error by more than 2%)

= $(.5)^{10}$ = 1/1024.

3-87 This exercise is solved in the same manner as 3-85 to arrive at the answer of $(.0478)^6$

3-89 If A is the event of the first one good and B that the second is good, then P(A and B) = P(A)P(B|A).
a. Since the first component was replaced, then P(B|A) = P(B) which is 1995000/2000000 = .9975. Then
P(A and B) = (.9975)(.9975) = .995.

b. Without replacement

P(A and B) = P(A)P(B|A)

= (1995000/2000000)(1994999/1999999) = .995.

3-91 The events are dependent. If A is the event that the first one is not stealing, B the second, C the third, and D the fourth, then

$$P(\text{A and B and C and D}) = \frac{34}{36} \bullet \frac{33}{35} \bullet \frac{32}{34} \bullet \frac{31}{33} = \frac{992}{1260} = .787.$$

3-93 a. We assume that the two events are independent. Since the probability of a person chosen at random having a supernatural experience is 1/17, then

P(two persons having such an experience) = $(1/17)^2$ = .0035.

b. P(two persons from Calif. having such an experience) = $(1/11)^2$ = .0083.

3-95 Assuming independence of the identifications, then the first woman could have selected any one of the men. Therefore,
P(the other 8 women select the same man) = $(1/5)^8$.

SOLUTIONS TO 3-4 B EXERCISES

3-99 a. There are 40 males among the 50 students, so the probability is 40/50 = .8.

b. There are 10 females so the probability is 10/50 = .2.

c. There were 20 yes answers among the students so the probability is 20/50 = .4.

d. There are 30 no answers so the probability is 30/50 = .6.

e. The probability of a yes answer is .4. Therefore, 40% of the male respondents would have answered yes. Then 40% of the 40 males would be 16 males. As a result, 16 of the 50 respondents were males who answered yes. That amounts to 32%.

f. see e.

g. Yes. The numbers will be as follows.

	Yes	No	
Male	16	24	40
Female	4	6	10
	20	30	

SOLUTIONS FOR SECTION 3.5 - COMPLEMENTARY EVENTS

3-101 The system is not defective, that is it works.

3-103 There are no "false" answers. All answers are "true".

3-105 If the event is D that the selection is defective, then P(D) = .1 = 5/50. The complementary event $\bar{D}$ is that the selection is not defective. Then

$$P(\bar{D}) = 1 - P(D) = 1 - (5/50) = 45/50 = .9.$$

3-107 If C is the event that the selected answer is correct, then P(E) = 1/5 = .2. The Complementary event $\bar{E}$ is that the selected answer is not correct. Then

$$P(\bar{E}) = 1 - P(E) = 1 - 1/5 = 4/5 = .8.$$

3-109 Let B be the event that the baby is a boy. Then P(B) = 1/2. The complementary event $\bar{B}$ is that the baby is not a boy. Then

$$P(\bar{B}) = 1 - P(B) = 1 - 1/2 = 1/2.$$

3-111 If W is the event that the student is a woman, then P(W) = 15/25 = 3/5 = .6 since there are 25 total students. The complementary event is that the selected student is not a woman. The probability of this complementary event is

$$P(\bar{W}) = 1 - P(W) = 1 - (15/25) = 10/25 = .4.$$

3-113 a. Since the probability of A is 2/7, then the probabitily of the complementary event is 5/7. Then the odds against A are

$$\frac{\frac{5}{7}}{\frac{2}{7}} = \frac{5}{2} \text{ expressed as 5:2.}$$

b. From a. the odds in favor of A are 2:5.

3-115 a. The probability of A is .4 or 2/5. Then the probability of $\bar{A}$ is 3.5. Using the methods of 3-113, the odds against A will be 3:2.

b. The odds in favor of A will be 2:3.

3-117 Given that the odds against A are 8:1, then

$$P(\bar{A}) = \frac{8}{8+1} = \frac{8}{9}.$$

Then

$$P(A) = \frac{1}{9}.$$

3-119 Given that the odds in favor of A are 3:7, then the probability of A is

$$P(A) = \frac{3}{3+7} = \frac{3}{10}.$$

3-121 Using complementary events, the probability of at least one employee 55 years old or older is 1 - the probability that neither is 55 or older. This is

$$\begin{aligned} P(\text{at least one}) &= 1 - P(\text{none 55 or older}) \\ &= 1 - P(\text{both are under 55}) \\ &= 1 - (.90)^2 = .19. \end{aligned}$$

3-123 Using complementary events, the probability of at least one is 1 - the probability of none indicating that a better job was a very important reason to go to college. That is

$$\begin{aligned} P(\text{at least one}) &= 1 - P(\text{none indicating jobs as very important}) \\ &= 1 - P(\text{all three went to college for other reasons}) \\ &= 1 - (.24)^3 = .986. \end{aligned}$$

3-125
$$\begin{aligned} P(\text{at least one pregnancy}) &= 1 - P(\text{no pregnancies among the 10}) \\ &= 1 - (.95)^{10} = 1 - .599 = .401. \end{aligned}$$

3-127
$$\begin{aligned} P(\text{the function is performed}) &= P(\text{an least one component works}) \\ &= 1 - P(\text{none work}) \\ &= 1 - P(\text{all fail}) \\ &= 1 - (.081)^4 = 1 - .000043 = .999957. \end{aligned}$$

3-129 The probability of any horse winning is 1/10 if all horses have the same chance of winning. The probability of the horse not winning is 9/10. The odds against the horse would be

$$\frac{P(\bar{A})}{P(A)} = \frac{\frac{9}{10}}{\frac{1}{10}} = \frac{9}{1} \text{ which we write as } 9:1.$$

3-131 The probability of guessing the first number correctly is 1/50. The probability of not guessing it is 49/50. So using the formula as in 3-129, the odds against it are 49:1.

3-133 The probability of "snake eyes" is 1/36. The probability of not getting "snake eyes" is 35/36. So using the formula as in 3-129, the odds against are 35:1. These would be the fair odds.

3-135 Using the probability of winning as 23/54, we proceed as in 3-133 to find the odds against as 31:23.

3-137
$$\begin{aligned} P(\text{at least one failure}) &= 1 - P(\text{no failures}) \\ &= 1 - P(\text{all 100 drivers pass}) \\ &= 1 - (.985)^{100} = 1 - .22 = .78. \end{aligned}$$

3-139 Because of the return on the wager, the odds were 18:2 against or 2:18 in favor of winning. Therefore the probability of winning is

$$\frac{2}{18 + 2} = \frac{2}{20} = .1.$$

3-141
$$\begin{aligned} P(\text{no misfire}) &= P(\text{at least one fail safe device worked}) \\ &= 1 - P(\text{all fail safe devices failed}) \\ &= 1 - (.105)(.2)(.001)(.115)(.34) \\ &= 1 - .0000008 = .9999992. \end{aligned}$$

3-143 This exercise is solved in the same manner as 3-131.

SOLUTIONS TO 3-5 B EXERCISES

3-145 a) $P(B) = 1 - P(A) = 1 - 1 = 0$.

b) $P(B) = 1 - P(A) = 1 - 0 = 1$.

c) $P(B) = 1 - P(A) \leq 1 - .7 = .3$.
It follows that $1 - P(A) \leq 1 - .7$ since we are subtracting a number $P(A)$ which is at least .7 from a number which is at most .3.

3-147 Since the odds against are 7:3 then the probability that the event did not occur is 7/10. The probability that the event did occur is 3/10. The probability that it occurred in all three separate trials is $(3/10)^3 = 27/1000$. The probability that the event did not occur in at least one of the trials is 973/1000. So the odds against the event occurring in all of three trials is 973:27.

SOLUTIONS FOR SECTION 3.6 - COUNTING

3-153 $7! = 7\bullet6\bullet5\bullet4\bullet3\bullet2\bullet1 = 5040$.

3-155 $$\frac{70!}{68!} = \frac{70\bullet69\bullet68\bullet\bullet\bullet1}{68\bullet67\bullet\bullet\bullet1} = 70\bullet69 = 4830.$$

3-157 $(9-3)! = 6! = 6 \bullet 5 \bullet 4 \bullet 3 \bullet 2 \bullet 1 = 720.$

3-159 By formula

$$_6P_2 = \frac{6!}{(6-2)!} = \frac{6!}{4!} = \frac{6 \bullet 5 \bullet 4 \bullet 3 \bullet 2 \bullet 1}{4 \bullet 3 \bullet 2 \bullet 1} = 6 \bullet 5 = 30.$$

3-161 By the combination formula

$$_{10}C_3 = \frac{10!}{3!(10-3)!} = \frac{10!}{3!7!} = \frac{10 \bullet 9 \bullet 8 \bullet 7 \bullet 6 \bullet 5 \bullet 4 \bullet 3 \bullet 2 \bullet 1}{3 \bullet 2 \bullet 1 \bullet 7 \bullet 6 \bullet 5 \bullet 4 \bullet 3 \bullet 2 \bullet 1} = 120.$$

3-163 $$_{52}C_2 = \frac{52!}{2!(52-2)!} = \frac{52 \bullet 51 \bullet 50!}{2 \bullet 1 \bullet 50!} = \frac{52 \bullet 51}{2} = 1326.$$

3-165 $$_nP_n = \frac{n!}{(n-n)!} = \frac{n!}{0!} = \frac{n!}{1} = n!.$$

3-167 $$_nC_0 = \frac{n!}{0!(n-0)!} = \frac{n!}{n!} = 1.$$

3-169 There are two groups by sex and 3 groups by income level so there are $2 \bullet 3 = 6$ total categories.

3-171 For arrangements, the order of selection matters. As a result, there are $_{10}P_4$ arrangements. By formula

$$_{10}P_4 = \frac{10!}{(10-4)!} = \frac{10!}{6!} = 10 \bullet 9 \bullet 8 \bullet 7 = 5040 \text{ arrangements.}$$

3-173 The number of combinations would not involve order but instead would be

$$_{22}C_{12} = \frac{22!}{(22-12)!12!} = \frac{22!}{10!12!}$$

$$= \frac{22 \bullet 21 \bullet 20 \bullet 19 \bullet 18 \bullet 17 \bullet 16 \bullet 15 \bullet 14 \bullet 13}{10 \bullet 9 \bullet 8 \bullet 7 \bullet 6 \bullet 5 \bullet 4 \bullet 3 \bullet 2 \bullet 1}$$

$$= 646646.$$

3-175 Since order matters, then the total number will be 10!.

3-177 a. Order of traversing the route matters, so there must be 7! = 5040 ways.

b. The probability of any one specific ordering is 1/(7!).

3-179 Each of the 9 digits of the social security number can be any of the 10 digits 0, 1, ••• 9. The total number would be 10^9.

3-181 Order of selection of the four individuals matters. Consequently there must be $_{16}P_4$ ways of selecting a slate.

$$_{16}P_4 = \frac{16!}{(16-4)!} = \frac{16!}{12!} = (16)(15)(14)(13) \text{ ways.}$$

3-183 There are 50 numbers to select from. Since the three numbers can be any of these, there are $(50)^3$ different sequences possible. They are in fact permutations.

3-185 There are a combination of 13 things taken two at a time ways to select a dessert. Therefore, there are

$10 \cdot 8 \cdot {}_{13}C_2 \cdot 3 = 18720$ arrangements.

3-187 a. We compute the combinations as follows.

$${}_{14}C_5 = \frac{14!}{5!(14 - 5)!} = \frac{(14)(13)(12)(11)(10)}{5 \cdot 4 \cdot 3 \cdot 2 \cdot 1}$$

$$= 14 \cdot 13 \cdot 11 = 2002.$$

b. There are 2002-650 = 1352 compatible combinations. Therefore the probability of selecting any one of these is 1352/2002 = .675.

3-189 a. Since order of selection is of no consequence the number would be the combination of 25 ponds selected 10 at a time. It is

$${}_{25}C_{10} = \frac{25!}{(25 - 10)!10!} = \frac{25!}{15!10!}$$

$$= \frac{25 \cdot 24 \cdot 23 \cdot 22 \cdot 21 \cdot 20 \cdot 19 \cdot 18 \cdot 17 \cdot 16}{10 \cdot 9 \cdot 8 \cdot 7 \cdot 6 \cdot 5 \cdot 4 \cdot 3 \cdot 2 \cdot 1}$$

$$= 5 \cdot 23 \cdot 22 \cdot 19 \cdot 2 \cdot 17 \cdot 2 = 3268760.$$

b. The probability of selecting any one particular sequence is 1/3268760.

3-191 The total number of routes would be 24!.

<u>SOLUTIONS TO 3-6 B EXERCISES</u>

3-193 a. There are a total of ${}_{60}C_5$ samples of with 5 members so the probability of any one being selected is $\frac{1}{{}_{60}C_5}$.

b. In general we would write the probability as

$\frac{1}{{}_NC_n}$ since there are ${}_NC_n$ different samples of size n.

3-195 Using the given formula

$$\frac{10!}{(10 - 3)!3!} = \frac{10!}{7!3!} = \frac{10 \cdot 9 \cdot 8}{3 \cdot 2} = 120.$$

3-197 Using the given formula for n = 50 we have

$$K = (50.5)\log(50) + .39908993 - .43429448(50)$$
$$= 85.797985 + .39908993 - 21.71472$$
$$= 64.482355.$$

Then $10^{64.482355} = 3.0364 \cdot 10^{64}$.

Using the calculator, we get $3.0414 \cdot 10^{64}$.

3-199 The first manager shakes hands with 4 others. Then the second shakes hands with 3 others. The handshake with the first was already counted. Proceeding like this, the total number of handshakes is 4 + 3 + 2 + 1 = 10.

SOLUTIONS FOR REVIEW EXERCISES

3-201 a. If A and B are independent, then P(A and B) = P(A)P(B) = (.2)(.4) = .08.

b. If A and B are mutually exclusive, then P(a or B) = P(A) + P(B) = .2 + .4 = .6.

c. $P(\bar{A})$ = 1 - P(A) = 1 - .2 = .8.

d. If P(A) = 2/10, then $P(\bar{A})$ = 8/10. The odds against A are .8/.2 = 4/1 which is expressed as 4:1.

e. P(C) = 1/(12 + 1) = 1/13.

3-203 Of the 6665 motion pictures, 2945 are rated R, so the probability is 2945/6665 = .442.

3-205 a. The probability of getting someone who wants to quit but doesn't is 3/4. The probability of getting someone who doesn't want to quit is 1/4. Therefore, the odds against finding someone who wants to quit but doesn't are 1:3.

b. Assuming independence of the two events, for A as smoker 1 and B as smoker 2, then
P(A and B wanting to quit) = P(A)P(B) = (3/4)(3/4) = 9/16.

3-207 One arrangement is in correct alphabetic order (assuming no two have the same last name). There are 7! arrangements. Thus the probability is 1/7! = 1/5040.

3-209 a. P(male or democrat)=P(male)+P(democrat)-P(male and democrat)
= 40/80 + 65/80 - 35/80 = 70/80.

b. P(male democrat) = P(democrat)P(male|democrat)
= (65/80)(35/65) = 35/80 = 7/16.

c. Since everyone in the sample is either democrat or republican, this probability is 1.0.

3-211 There are 4 patients and one of them has type O blood. Therefore the probability is 1/4 = .25.

3-213 P(healthy or group O)=P(healthy)+P(group O)-P(healthy and group O)
= 39/60 + 30/60 - 20/60 = 49/60.
The reasons for the above are that there are 39 healthy people (adding across) there are 30 group O people (adding down) and 20 people who are healthy and of group O blood.

3-215 The probability of not winning is 6135755/6135756. As a result the odds against winning are 6135755:1.

3-217 a) 8! = 8•7•6•5•4•3•2•1 = 40320.

b) $P_{8\,6} = \frac{8!}{(8-6)!} = \frac{8!}{2!} = \frac{8\bullet7\bullet6\bullet5\bullet4\bullet3\bullet2\bullet1}{2\bullet1} = 20160.$

c) $C_{10\,8} = \frac{10!}{(10-8)!8!} = \frac{10!}{2!8!} = \frac{10\bullet9}{2\bullet1} = 45.$

d) $C_{80\,78} = \frac{80!}{2!78!} = \frac{80\bullet79}{2} = (40)(79) = 3160.$

3-219 There are 20 people at the hearing and 8 are opposed. Therefore, the probability is 8/20.

3-221 There are 5! = 120 ways of ordering 5 events. As a result the probability of selecting any one is 1/120.

3-223 Since order of selection and arrangement matters, this is a permutation. The number is

$${}_8P_3 = \frac{8!}{(8-3)!} = \frac{8!}{5!} = \frac{8\bullet7\bullet6\bullet5!}{5!} = 8\bullet7\bullet6 = 336.$$

3-225 Using the general formula and A for the first selected patient and B for the second,

$$P(\text{A and B had unfavorable reactions}) = P(A)P(B|A) = \frac{35}{60}\bullet\frac{34}{59} = \frac{1190}{3540}.$$

3-227 Let A be the event that at least one improves. Then $\bar{A}$ is the event that all do not improve. Using the formula for complementary events,

$$P(A) = 1 - P(\bar{A}) = 1 - (.40)^{10} = 1 - .0001 = .9999.$$

3-229 The total number of configurations is 6•4•8•4 = 768. event of none

3-231 Since the order of arrangement matters, then this is the permutation of 12 songs selected 4 at a time. That is

$${}_{12}P_4 = \frac{12!}{(12-4)!} = \frac{12!}{8!} = \frac{12\bullet11\bullet10\bullet9\bullet8!}{8!} = 11880.$$

3-233 Since the odds against the event are 7:1, then the probability of the event is 1/(1 + 7) = 1/8. The probability of the event in two separate trials will be (1/8)(1/8) = 1/64. The probability of the event not so occuring is 63/64. Therefore, the odds against the event occuring in two separate and independent trials is 63:1.

3-235 Since there are 6 departure cities and 4 arrival cities, there are a total of 6•4 = 24 possible routes.

CHAPTER 4. PROBABILITY DISTRIBUTIONS

SOLUTIONS TO SECTION 4.2 - RANDOM VARIABLES

4-1 The variables of temperature, weight, and time are examples of continuous random variables. They are a,c, and d. The others are discrete.

4-3 This is a probability distribution because the sum of the probabilities of the random variables is 1 and all $P(x) \geq 0$.

4-5 This is not a probability distribution, because the sum of the probabilities of the random variables is .78, which is less than 1.

4-7 This is a probability distribution, because each probability is between 0 and 1 and the sum of the probabilities of the random variables is 1 as seen in the table.

x	P(x)
.1	.1
.3	.3
.6	.6

4-9 Using the formula to determine probabilities of the various events, we will have a table as given below. Because each probability is between 0 and 1 and the sum of the probabilities is 1, this is a probability distribution.

x	P(x)
.5	0
.6	.1
.7	.2
.8	.3
.9	.4

4-11 Using the formula to determine the various probabilities, we have

$$p(0) = \frac{1}{2(2)!0!},\ p(1) = \frac{1}{2(1)!1!},\ \text{and } p(2) = \frac{1}{2(0)!2!}.$$

Then $P(0) = 1/4$, $p(1) = 1/2$, and $p(2) = 1/4$.

These are all between 0 and 1 and are therefore probabilities. Their sum is 1. As a result this is a probability distribution.

4-13 a. The values of the random variables correspond to the number of patients cured. Consequently, there would be the values 0, 1, 2, and 3.

b. As stated in the problem, $p(0) = .125$, $p(1) = .375$, $p(2) = .375$, and $p(3) = .125$.

c. The table would be:

x = cures	P(x)
0	.125
1	.375
2	.375
3	.125

d. Same as part b.

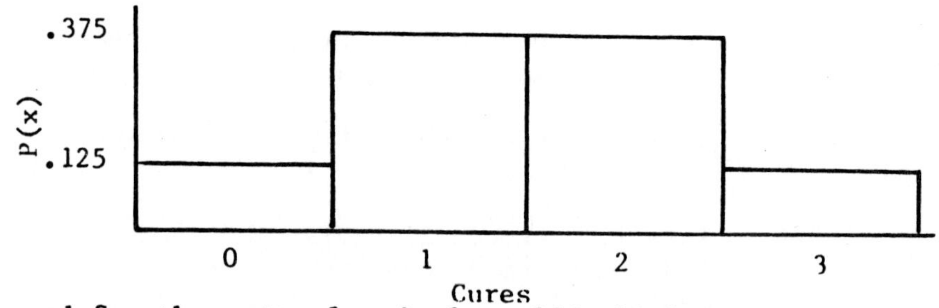

e. The area of the first and fourth rectangles is 1 x .125 which is .125, the probabilities of those two events. The area of the second and third rectangles is 1 x .375 = .375, the probabilities of those two events.

4-15 a. The values of the random variable would only be 0, 1, or 2.

b. Since each one is equally likely, then p(0) = 1/3, p(1) = 1/3, and p(2) = 1/3.

c. The table would be:

x	p(x)
0	1/3
1	1/3
2	1/3

d.

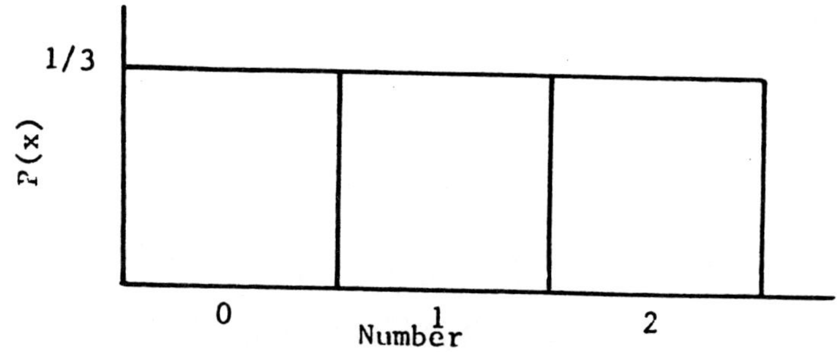

e. Each rectangular region in d above is of width 1 and height 1/3. Consequently, its area is 1/3 x 1 = 1/3, the probability of the event.

4-17 a. The number of girls in a family of 3 children would be 0, 1, 2, or 3.

b. By looking at the equally likely events of the sample space listed in the hint, we see that p(0) = 1/8, p(1) = 3/8, p(2) = 3/8, and p(3) = 1/8.

c. The summarizing table is below. All probabilities are between 0 and 1, and their sum is 1, so it so it is a probability distribution.

x	p(x)
0	1/8
1	3/8
2	3/8
3	1/8

d.

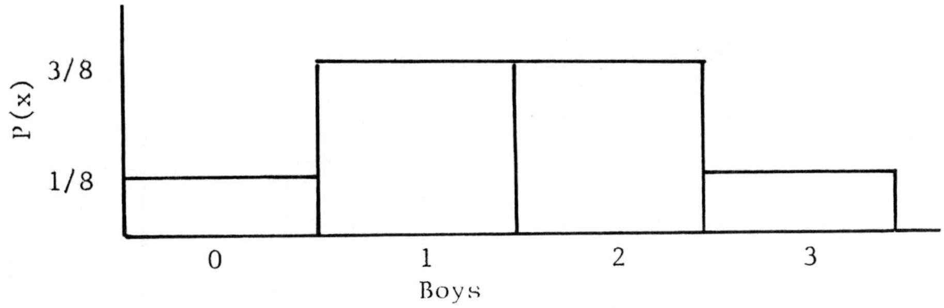

e. The first and fourth rectangles have area 1/8, the second and third rectangles have area 3/8.

4-19 a. The possible values taken on by the random variable would be the numbers 2, 3, 4, 5, 6, 7, 8, 9, 10, 11, and 12.

b. We list the values in the sample space along with the associated value of the random variable in the following table.

Two Dice Outcome	Random Variable	Two Dice Outcome	Random Variable
1,1	2	2,1	3
1,2	3	2,2	4
1,3	4	2,3	5
1,4	5	2,4	6
1,5	6	2,5	7
1,6	7	2,6	8
3,1	4	4,1	5
3,2	5	4,2	6
3,3	6	4,3	7
3,4	7	4,4	8
3,5	8	4,5	9
3,6	9	4,6	10
5,1	6	6,1	7
5,2	7	6,2	8
5,3	8	6,3	9
5,4	9	6,4	10
5,5	10	6,5	11
5,6	11	6,6	12

c. We summarize the value of the random variable and its associated probability in the following table:

Random Variable X	P(x)
2	1/36
3	2/36
4	3/36
5	4/36
6	5/36
7	6/36
8	5/36
9	4/36
10	3/36
11	2/36
12	1/36

d. The histogram would be:

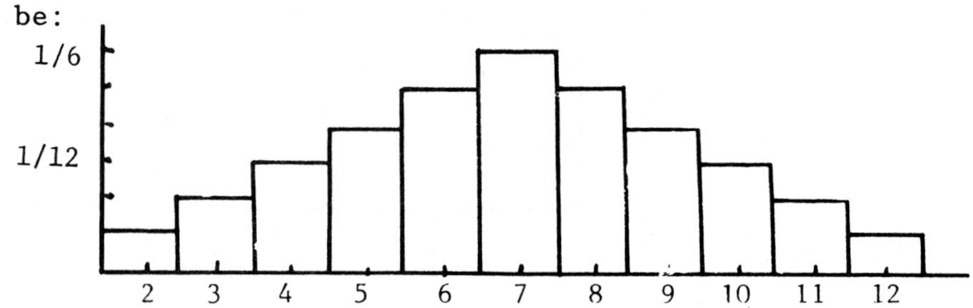

e. Each rectangle is of width 1 and height equal to the probability of the random variable. Therefore, the area is the probability of the random variable.

SOLUTIONS TO 4-2 B EXERCISES

4-21 It is for if we look at $.4(.6)^{x-1}$, this will be positive for any value of x. In addition

$$P(x) = .4(1 + .6 + .6^2 + .6^3 + \cdots)$$
$$= .4(\frac{1}{1-.6}) = .4\frac{1}{.4} = 1 \quad \text{as a geometric series.}$$

4-23 We first compute the probability of no acceptable ones in 4 chips when the probability of any one not being a defective is .35. Then

$$p(0) = (.65)^4 = .179.$$

because they are "and" probabilities to be multiplied. The other probabilities would be given by

$$p(4) = (.35)^4 = .015.$$

Then, $p(3) = 1 - .015 - .179 - .384 - .311 = .111$

SOLUTIONS TO SECTION 4.3 - MEAN, VARIANCE, AND EXPECTATION

4-25 We form a table as was done in the example in the text.

	x	P(x)	x•P(X)	x^2	x^2•P(x)
	0	.2	0	0	0
	1	.5	.5	1	.5
	2	.3	.6	4	1.2
Totals			1.1		1.7

Then the mean $\mu = 1.1$ and the variance $\sigma^2 = \sum x^2 \bullet P(x) - \mu^2$

$$= 1.7 - 1.21 = .49.$$

The standard deviation $\sigma = \sqrt{.49} = .7.$

4-27 Forming our table to aid computations, we have the following:

x	P(x)	x•P(x)	x^2	x^2•P(x)
1	.15	.15	1	.15
2	.45	.90	4	1.80
3	.35	1.05	9	3.15
4	.05	.20	16	.80
Totals		2.30		5.90

The mean μ = 2.30 and $\sigma^2 = 5.9 - (2.3)^2 = .61$.

The standard deviation $\sigma = \sqrt{.61} = .781$.

4-29 We form our table as was done above.

x	P(x)	x•P(x)	x^2	x^2•P(x)
5	1/4	5/4	25	25/4
10	1/4	10/4	100	100/4
20	1/4	20/4	400	400/4
50	1/4	50/4	2500	2500/4
Totals		85/4 = 21.25		3025/4 = 756.25

Then μ = 21.25 and the variance $\sigma^2 = 756.25 - (21.25)^2 = 304.69$.

The standard deviation $\sigma = \sqrt{304.69} = 17.46$.

4-31 Forming the table we have the following:

x	P(x)	x•P(x)	x^2	x^2•P(x)
2	4/30	8/30	4	16/30
4	6/30	24/30	16	96/30
6	10/30	60/30	36	360/30
8	6/30	48/30	64	384/30
10	4/30	40/30	100	400/30
Totals		180/30 = 6		1256/30 = 41.9

The mean $\mu = \sum x \bullet P(x) = 6$, while the variance $\sigma^2 = \sum x^2 \bullet P(x) - \mu^2 =$ $41.9 - 36 = 5.9$. The standard deviation is $\sigma = \sqrt{5.9} = 2.4$.

4-33 The table to aid computation is given below.

x	P(x)	x•P(x)	x^2	x^2•P(x)
0	.026	0	0	0
1	.154	.154	1	.154
2	.346	.692	4	1.384
3	.346	1.038	9	3.114
4	.130	.52	16	2.08
Totals		2.404		6.732

Then μ = 2.404 and $\sigma^2 = 6.732 - (2.404)^2 = .952$.

The standard deviation is $\sigma = \sqrt{.952} = .976$.

4-35 The table for computations is as follows:

x	P(x)	x•P(x)	x^2	x^2•P(x)
0	.002	0	0	0
1	.020	.020	1	.020
2	.095	.190	4	.38
3	.235	.705	9	2.115
4	.328	1.312	16	5.248
5	.244	1.220	25	6.1
6	.075	.450	36	2.7
Totals		3.897		16.563

The mean $\mu = \sum x \bullet P(x) = 3.897$ and the variance $\sigma^2 = 16.563 - (3.897)^2 = 1.376$. The standard deviation is $\sigma = \sqrt{1.376} = 1.17$.

4-37 We can form our table as follows:

x	P(x)	x•P(x)	x^2	x^2•P(x)
0	.65	0	0	0
1	.3	.3	1	.3
2	.05	.10	4	.2
Totals		.4		.5

Then the mean number of accidents on a given weekday is $\mu = \sum xP(x) = .4$. The variance is $\sigma^2 = .5 - (.4)^2 = .5 - .16 = .34$. The standard deviation is $\sigma = .583$.

4-39 Forming the table of numbers of World Series games we have

x	P(x)	x•P(x)	x^2	x^2•P(x)
4	.12	.48	16	1.92
5	.253	1.265	25	6.325
6	.217	1.302	36	7.812
7	.410	2.87	49	20.09
Totals		5.917		36.147

As before, $\mu = 5.917$ and $\sigma^2 = 36.147 - (5.917)^2 = 1.14$ and $\sigma = 1.07$.

4-41 Your expected value is $E(x) = x_1P(x_1) + x_2P(x_2)$. Then

$$E(x) = \frac{1}{4}(\$500) + \frac{3}{4}(\$200) = \$275.$$

4-43 The formula for expected value is

$$E(x) = \sum xP(x) = \frac{1}{10}(\$200) + \frac{3}{10}(-\$300) + \frac{6}{10}(0)$$

$$= -\$70.$$

4-45 As in the above

$$E(x) = \sum xP(x) = .7(\$175,000) + .3(0) = \$122,500.$$

4-47 For the expected return

$$E(x) = \sum xP(x) = .6(\$200) + .4(-\$150) = \$60.$$

SOLUTIONS TO 4-3 B EXERCISES

4-49 a. In order of birth the possible outcomes are:

bbbbb	bbbgg	gggbb	ggggg
bbbbg	bbgbg	ggbgb	ggggb
bbbgb	bgbbg	gbggb	gggbg
bbgbb	gbbbg	bgggb	ggbgg
bgbbb	bbggb	ggbbg	gbggg
gbbbb	bgbgb	gbgbg	bgggg
	gbbgb	bggbg	
	bggbb	gbbgg	
	gbgbb	bgbgg	
	ggbbb	bbggg	

b. Using a to establish probabilities:

x	P(x)
0	1/32
1	5/32
2	10/32
3	10/32
4	5/32
5	1/32

c.

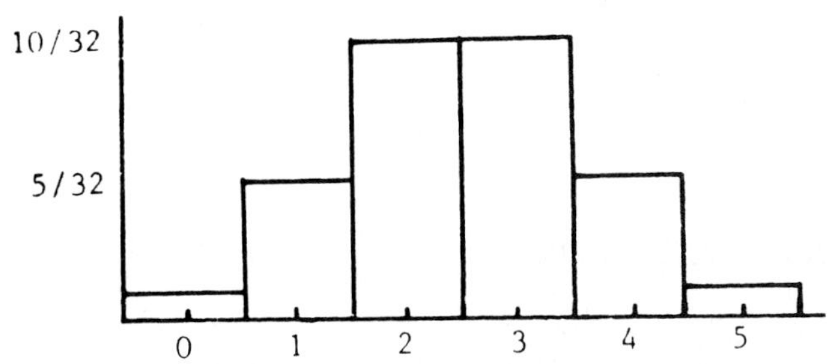

d,e,f. Using the table to guide computations:

x	P(x)	x•P(x)	x^2	x^2•P(x)
0	1/32	0	0	0
1	5/32	5/32	1	5/32
2	10/32	20/32	4	40/32
3	10/32	30/32	9	90/32
4	5/32	20/32	16	80/32
5	1/32	5/32	25	25/32
Totals		80/32 = 2.5		240/32 = 7.5

Then $\mu = 2.5$, $\sigma^2 = 7.5 - (2.5)^2 = 1.25$. The standard deviation is $\sigma = 1.12$.

g. See part c.

4-51 a. The mean would be $\mu = \sum xP(x)$

$$= 1 \bullet \frac{1}{n} + 2 \bullet \frac{1}{n} + \bullet\bullet\bullet\ n \bullet \frac{1}{n} = \frac{1}{n}(1 + 2 + \bullet\bullet\bullet + n)$$

$$= \frac{1}{n}\left(\frac{n(n+1)}{2}\right) = \frac{n+1}{2}$$

using the formula of the hint.

b. $\sum x^2P(x) = 1^2 \bullet \frac{1}{n} + 2^2 \bullet \frac{1}{n} + \bullet\bullet\bullet + n^2 \bullet \frac{1}{n}$

$$= \frac{1}{n}(1^2 + 2^2 + \bullet\bullet\bullet + n^2)$$

$$= \frac{1}{n}\left(\frac{n(n+1)(2n+1)}{6}\right) = \frac{(n+1)(2n+1)}{6}.$$

In the third step we use the formula of the hint for the sum of the first n squares. Then

$$\sigma^2 = \sum x^2P(x) - \mu^2 = \frac{(n+1)(2n+1)}{6} - \left(\frac{n+1}{2}\right)^2$$

$$= (n+1)\left(\frac{2n+1}{6} - \frac{n+1}{4}\right)$$

$$= (n+1)\left(\frac{4n+2}{12} - \frac{3n+3}{12}\right)$$

$$= (n+1)\left(\frac{n-1}{12}\right) = \frac{n^2-1}{12}.$$

<u>SOLUTIONS TO SECTION 4.4 BINOMIAL EXPERIMENTS</u>

4-57 a. Using the formula

$$\frac{5!}{(5-3)!3!} = \frac{5!}{2!3!} = \frac{5\bullet4\bullet3\bullet2\bullet1}{2\bullet1\bullet3\bullet2\bullet1} = 10.$$

c. $\frac{8!}{(8-7)!7!} = \frac{8!}{1!7!} = \frac{8\bullet7\bullet6\bullet5\bullet4\bullet3\bullet2\bullet1}{1\bullet7\bullet6\bullet5\bullet4\bullet3\bullet2\bullet1} = 8.$

e. $\frac{8!}{0!8!} = 1$ since $0! = 1$.

4-59 a. $P(3) = \frac{5!}{(5-3)!3!}\left[\frac{1}{4}\right]^3\left[\frac{3}{4}\right]^{(5-3)} = 10\bullet\frac{9}{1024} = \frac{90}{1024}$

b. $P(2) = \frac{4!}{(4-2)!2!}\left[\frac{1}{3}\right]^2\left[\frac{2}{3}\right]^{(4-2)} = 6\bullet\frac{4}{81} = \frac{8}{27}.$

c. $P(1) = \frac{5!}{(5-1)!1!}\left[\frac{2}{3}\right]^1\left[\frac{1}{3}\right]^{(5-1)} = 5\bullet \frac{2}{243} = \frac{10}{243}$.

4-61 Using the binomial formula with n = 10, x = 4, p = 1/2 and 1 - p = q = 1/2, we have

$$p(x = 4) = \frac{10!}{(10-4)!4!}\left[\frac{1}{2}\right]^4\left[\frac{1}{2}\right]^{(10-4)}$$

$$= \frac{10\bullet 9\bullet 8\bullet 7\bullet (6!)}{6!\bullet 4\bullet 3\bullet 2\bullet 1}\bullet \left(\frac{1}{2}\right)^4\left(\frac{1}{2}\right)^6 = \frac{210}{2^{10}} = \frac{210}{1024} = .205.$$

4-63 The number n in the sample is 5. The probability of success is p = .6 and then q = .4. We are looking for the probability that x = 3. Then

$$p(3) = \frac{5!}{2!3!}(.6)^3(.4)^2 = \frac{5\bullet 4}{2!}\bullet (.6)^3(.4)^2 = .3456.$$

4-65 Since 95% survive, the probability of surviving is p = .95 while q = .05. From the problem n = 20 and x = 16. Then, using Table A-1 the probability is .013.

4-67 Here p = .4, q = .6, n = 6. Then again using Table A-1,
a. P(0) = .047
b. P(3) = .276
c. P(at least 3) = P(3) + P(4) + P(5) + P(6)
= .276 + .138 + .037 + .004 = .455

4-69 Using the data of n = 15, x = 13, p = .93 and q = .07, we would have

$$p(13) = \frac{15!}{13!2!}(.93)^{13}(.07)^2 = 105(.389)(.005) = .20.$$

4-71 We have p = .15 and q = .85. For n = 60 and x = 12 we want
$$p(12) = \frac{60!}{12!48!}(.15)^{12}(.85)^{48} = .0743.$$

4-73 The probability of a catalytic converter being removed is .044. Then q = .956 and n = 30.
a. P(0 converters removed) = $.956^{50}$ = .105.

b. P(1) = $50\bullet .956^{49}.044$ = .243.

c. $P(2) = \frac{50!}{2!48!}(.956)^{48}(.044)^2 = \frac{50\bullet 49}{2}(.115)(.0019) = .268.$

4-75 Since p = .01, then q = .99. The exact value is

$$p(2) = \frac{3!}{2!1!}(.01)^2(.99)^1 = 3(.0001)(.99) = .000297.$$

4-77 As in table 4-4, we need values for x. In this case x is the number of successes in 3 trials. Therefore, x can only take on the values of 0, 1, 2, 3. Here, p = 1/3.

x	p(x)
0	.296
1	.444
2	.222
3	.038

$$p(0) = \frac{3!}{0!3!} (\frac{1}{3})^0 (\frac{2}{3})^3 = .296.$$

$$p(1) = \frac{3!}{1!2!} (\frac{1}{3})^1 (\frac{2}{3})^2 = .444.$$

$$p(2) = \frac{3!}{2!1!} (\frac{1}{3})^2 (\frac{2}{3})^1 = .222.$$

4-79 Here we have p = .98, q = .02. The probability of 74 out of 75 passing is

$$P(74) = \frac{75!}{74!1!} (.98)^{74}(.02)^1 = .336.$$

4-81 This exercise is solved in a manner similar to 4.79. For the probability of fewer than two, we will need to compute using the formula, the probability of 0 and of 1.

4-83 The probability of more than 8 is the probability of 9 plus the probability of 10. These two are computed using the binomial formula as in 4-79. The two probabilities are then added to get the final answer.

SOLUTIONS TO 4-4 B EXERCISES

4-85 The first defect as the seventh component tested means that the first six are acceptable. Using the formula,
$P(7) = (.8)^6(.2) = .052$

4-87 We use the formula to obtain

$$P(4) = \frac{7!}{(7-4)!4!} \bullet \frac{3!}{(3-5+4)!(5-4)!} \div \frac{10!}{(10-5)!5!}$$
$$= \frac{7\bullet6\bullet5}{3\bullet2} \bullet \frac{3\bullet2}{2} \bullet \frac{5\bullet4\bullet3\bullet2}{10\bullet9\bullet8\bullet7\bullet6} = \frac{5}{12} = .417.$$

SOLUTIONS TO SECTION 4.5 MEAN AND STANDARD DEVIATION FOR THE BINOMIAL DISTRIBUTION

4-89 Using the formulas we have μ = np = (64)(.5) = 32 and σ^2 = npq = 64(.5)(.5) = 16.
The standard deviation is $\sigma = \sqrt{16} = 4$.

4-91 Using the formulas μ = np = 8(.6) = 4.8 while the variance is σ^2 = npq = 8(.6)(.4) = 1.92. Then σ = 1.39.

4-93 Using the formula, μ = np = 36(.25) = 9. The variance is σ^2 = npq = 36(.25)(.75) = 6.75. Then the square root of the variance is the standard deviation which is σ = 2.6.

4-95 See 4-93 for the method.

4-97 By the formulas μ = np = (16)(1/5) = 3.2 and σ^2 = (18)(1/5)(4/5) = 2.56 Then the standard deviation is $\sigma = \sqrt{2.56}$ = 1.6.

4-99 In this problem n = 253 and p = 2/3. Then μ = np = 253• (2/3) = 168.67 and σ^2 = 253(2/3)(1/3) = 56.22.Then the standard deviation is $\sigma = \sqrt{56.22}$ = 7.5.

4-101 Since p = .5 for the probability of a correct guess, then μ = np = 50(.5) = 25. Also σ^2 = (50)(.5)(.5) = 12.5. The standard deviation is $\sigma = \sqrt{12.5}$= 3.54.

4-103 We have n = 4 while p = .5 the probability of a girl. Then μ = (4)(.5) = 2 and σ^2 = npq =(4)(.5)(.5) = 1. Then $\sigma = \sqrt{1}$ = 1.

4-105 Here we have p = .48 and n = 64. Then μ = np = (64)(.48) = 30.7 and the variance is σ^2 = (64)(.48)(.52) = 16. As a result, σ = 4.

4-107 We have p = 2945/6665 = .44 and n = 20. As a result, μ = np = (.44)(20) = 8.8. The variance is σ^2 = npq = (20)(.44)(.56) = 4.93. Then σ = 2.22.

4-109 Because p = .47 and n = 200, we have μ = np = (200)(.47) = 94 and σ^2 = npq = (200)(.47)(.53) = 49.8. Consequently σ = 7.06.

4-111 In this problem p = .37 and n = 12. Hence μ = (.37)(12) = 4.44. In addition, σ^2 = npq = (12)(.37)(.63) = 2.8. Then the standard deviation is 1.67.

4-113 Solved by the same method as 4-111.

4-115 Solved in the same manner as 4-111.

4-117 Since p = .138 for officers, then μ = np = (50)(.138) = 6.9. Then σ^2 = (50)(.138)(.862) = 5.95. Consequently σ = 2.44.

4-119 Solved in the same manner as 4-117.

SOLUTIONS TO 4-5 B EXERCISES

4-121 We have μ = (.5)(1000) = 500 and $\sigma = \sqrt{(.5)(.5)(1000)}$ = 15.8. Then, 450 girls is not in the interval from $\mu-2\sigma$ = 468.4 to $\mu+2\sigma$ = 531.6. Hence, the event would be unusual.

4-123 Since n = 10 and p = .05 for defectives, then μ = np = .5 and $\sigma^2 = \sqrt{(.05)(.95)(10)}$ = .69. Then $\mu - 2\sigma$ = .5 - 1.38 = -.88 while

$\mu + 2\sigma$ = .5 + 1.38 = 1.88. The observed number of defectives of 2 is not in this interval, so it would be considered unusual.

SOLUTIONS TO SECTION 4.6 DISTRIBUTION SHAPES

4-125 a. Since 20 of the 100 degree range is greater than 480 degrees, then the probability of more than 480 will be 20/100 = .2.

b. There are 80 degrees between 410 and 490. Therefore, the probability will be 80/100 = .8.

c. There are 50 degrees between 450 and 500, so the probability is 50/100 = .5.

d. There are 65 degrees between 400 and 465, so the probability is 65/100 = .65.

4-127 The range of the data is 30 seconds.
a. There are 20 seconds in the range more than 270 so the probability is 20/30 = .67.

b. There are 22 seconds in the range between 265 and 287, so the probability is 22/30 = .733.

c. There are 8 seconds in the range less than 268, so the probability is 8/30 = .267.

e. The mean of a uniform distribution is is midpoint of the range. The probability of being less than that is .5.

4-129 Since the range is 10, the data runs uniformly from 30 to 40.

a. There are 5 of the 10 mile range less than 35, so the probability is .5.

b. There are 8 of the units in the interval greater than 32, so the probability is .8.

c. The interval is 6 mph of the range of 10. The probability is then 6/10 = .6.

d. The median and the mean are the same in a uniform distribution, so the probability is .5.

e. The interval from 35 to 45 intersects our interval in only a 5 mph interval. Therefore the probability is .5.

4-131 Solved in the same manner as 4-129.

4-133 We draw the diagram as follows.

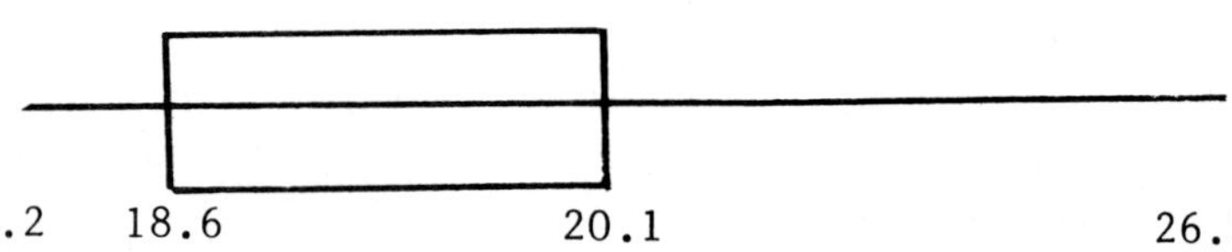

The hinge 18.6 which is the median of the lower half of the data 17.2, 17.9, 18.6, 18.8, and 19.3. The upper hinge is 20.1 as the median of the data 19.3, 20.0, 20.1, 23.4, and 26.3.

4-135 As in 4-133 we draw our figure using the following points. We find the median of the lower half of the data .02, .04, .08, .08, .09, and .095. That median is .08. The upper hinge of .11 is the median of .095, .1, .1, .12, .13, and .19.

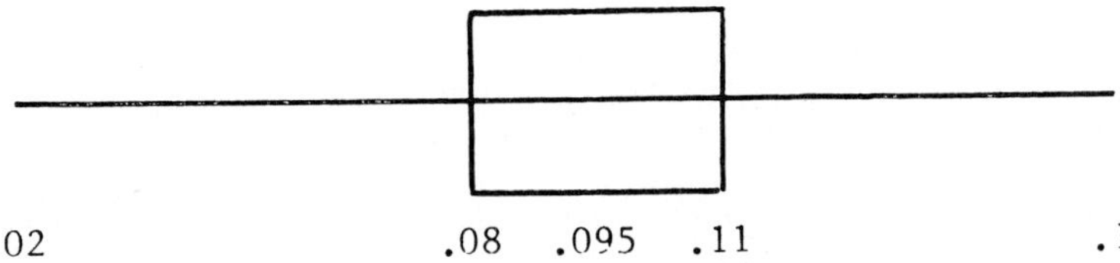

4-137 The minimum is 130 and the maximum is 198. Ordering the data gives a median of 171.5. The two hinges are computed as above to be 159 and 181 respectively. The diagram is as follows.

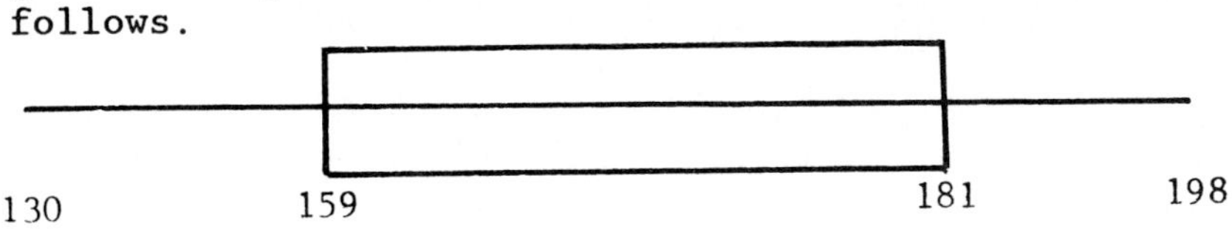

4-139 Using the methods described above in 4-135 to find the points, the diagram is as follows.

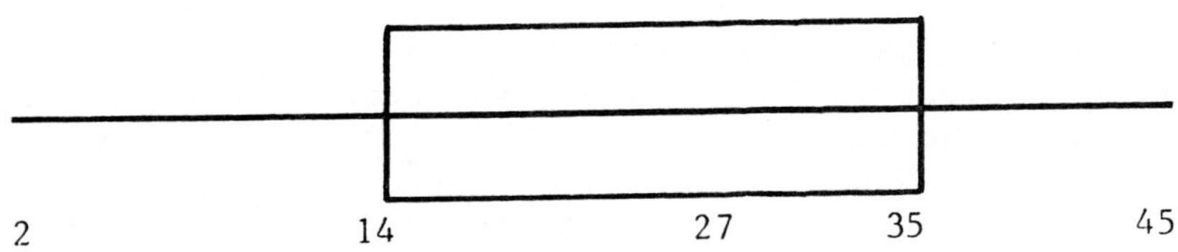

SOLUTIONS TO 4-6 B EXERCISES

4-141 a. The area A less than 3 has base 3 and height 60. Then the area of A is (1/2)(3)(60) = 90. Since the whole area is 250, the probability is 90/250 = .36.

b. The area of the segment is the area of the larger segment minus the area of the smaller. This is (1/2)(4)(80) - (1/2)(2)(40) = 120. As a result the probability is 120/250 = .48.

SOLUTIONS TO REVIEW EXERCISES

4-145 a. For all five to be democrats we will have a product of probabilities. Consequently the probability will be $(.4)^5 = .01$.

b. We can use either table A2 or the formula for binomial probabilities. From the table with $p = .4$, $q = .6$, $n = 5$, and $x = 3$, the probability is .230.

c. The mean is $\mu = np = 5(.4) = 2$.

d. The variance is $\sigma^2 = npq = 5(.4)(.6) = 1.2$.

e. The standard deviation is the square root of the variance, which is 1.095.

4-147 The whole interval is of length 40. The interval of interest is from 50 to 75 and is of length 25. As a result the probability is 25/40 = 5/8.

4-149 It is not a probability distribution because the sum of the probabilities of the events is 1.2 which is greater than 1.0.

4-151 a. This is a binomial random variable with $n = 20$, $p = .05$ and $x = 3$. Then using the binomial table A-1, we find the probability to be .06.

b. The mean $\mu = np = (20)(.05) = 1$.

c. The variance is $\sigma = npq = (20)(.05)(.95) = .95$. The standard deviation is the square root of this which is .975.

4-153 This is a binomial random variable.

a. We are seeking the probability of x = 1 divorced person out of a sample of n = 10. In this case p = 5.3/1000 = .0053. Using the formula we have

$$P(1) = \frac{10!}{1!9!}(.0053)^1(.9947)^9 = .051.$$

b. The variance is $\sigma^2 = npq = 1000(.0053)(.9947)^9 = 5$. The standard deviation is the square root of this, which is 2.3.

4-155 For the box and whisker diagram we reorder the data in increasing order of magnitude:

142 142 153 155 163 165 167 171

The median of these is 159. The median of the lower half of the data including 159 is 153. The median of the upper half of the data including 159 is 165. These are the two hinges. The diagram is as follows.

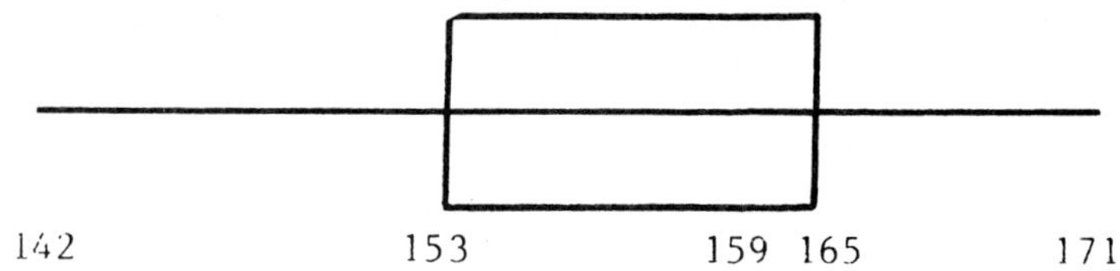

4-157 The number of offspring here is a binomial random variable.

a. Using the table of binomial probabilities, we find that the probability of 8 is .196.

b. The mean is $\mu = np = 16(.5) = 8$.

c. The variance is $\sigma^2 = npq = 16(.5)(.5) = 4$.

4-159 The whole interval is of length 2 volts. The interval below 8.55 volts is of length 1.55 volts. As a result the probability of a sample lying in that interval is 1.55/2 = .775.

4-161 It is not a probability distribution because P(-1)+P(0)+P(1)+P(2)= 1/5 + 21/5 + 3/5 = 6/5. The sum of the probabilities is then greater than 1.0.

4-163 a. The table of values of the random variable and their associated probabilities is given below.

x	P(x)
1	.2
2	.2
3	.2
4	.2
5	.2

b. The mean is 1(.2) + 2(.2) + 3(.2) + 4(.2) + 5(.2) = 3.

c. Using the formula for variance we need $\sum x^2P(x)$. It is given by $1^2(.2) + 2^2(.2) + 3^2(.2) + 4^2(.2) + 5^2(.2) = 11$. Then $\sigma^2 = \sum x^2P(x) - \mu^2 = 11 - 9 = 2$.

4-165 a. The probability of at least one component costing less than 50 cents is 1 - the probability of none costing less than 50 cents. The probability of none in a group of 8 will be $(13/15)^8$, since the probability of getting one in one selection is 13/15 and we make 8 selections. Then the probability of at least one is $1 - (13/15)^8 = .682$.

b. The mean is $\mu = np = 8(2/15) = 1.07$.

c. The standard deviation is $\sigma = \sqrt{8(2/15)(13/15)} = .96$.

4-167 The median is 77.5. As a result the two hinges are medians of the lower and upper half of the data. The values are 72 and 82 respectively. The diagram is given below.

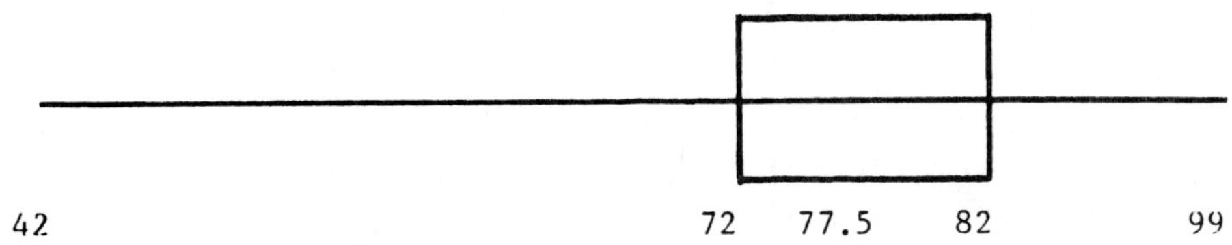

CHAPTER 5. NORMAL PROBABILITY DISTRIBUTIONS

SOLUTIONS TO SECTION 5.2 - THE STANDARD NORMAL DISTRIBUTION

5-1 The probability of lying between 0 and a positive number comes directly from the table as .0987.

5-3 As in 5-1, we have from the table .4332.

5-5 The probability of lying between -1.0 and 0 is the same as the probability of lying between 0 and +1.0 which is .3413. As areas this is seen in the figure below.

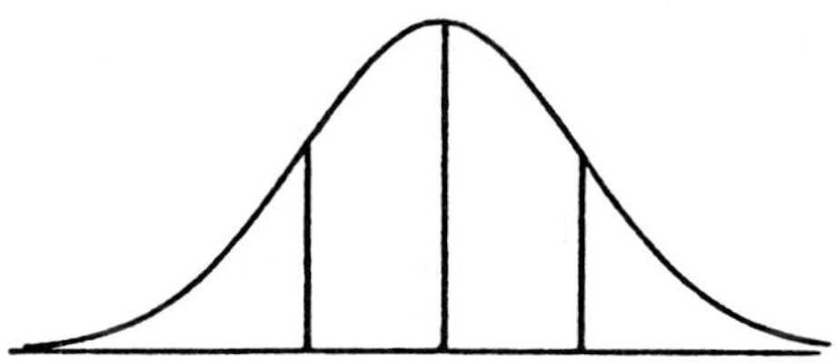

5-7 As in 5-5, this is the same as the probability that z lies between 0 and +1.75. This probability is .4599.

5-9 The probability of a thermometer reading greater than 1.0 plus the probability of one reading between 0 and 1 is .5. This situation is seen in the figure below. Therefore, since the probability from the table of a z value of 1.0 is .3413, the probability of more than 1 is .5 - .3413 = .1587.

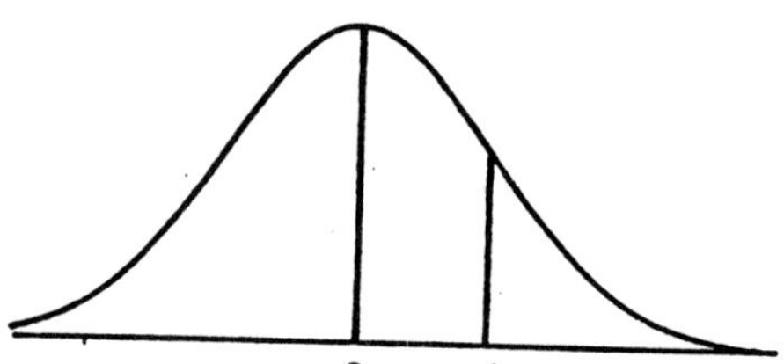

5-11 As in 5-9, the probability of greater than 1.83 is .5 minus the probability of lying between 0 and 1.83. From the normal table this is .4664. Then the probability of lying more than 1.83 is .5 - .4664 = .0336.

5-13 The probability of a reading less than -1. plus the probability of a reading between -1. and 0. is .5. The probability of a reading between -1. and 0 is .3413 (exercise 5-5). Then the probability of a reading less than -1.0 is .5 - .3413 = .1587.

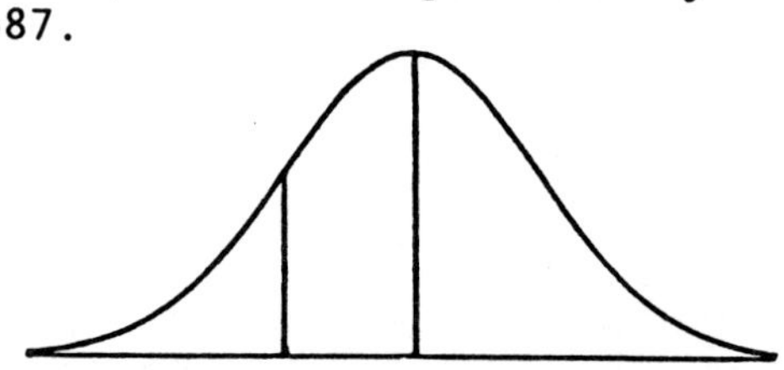

5-15 As is 5-13, this is .5 minus the probability of lying between 0 and -.91. This is .5 - .3186 = .1814.

5-17 The probability of a reading greater than -1. is the probability of a reading between -1. and 0 plus the probability of a reading more than 0.(See the figure below) The resultant probability is .5 + .3413 = .8413.

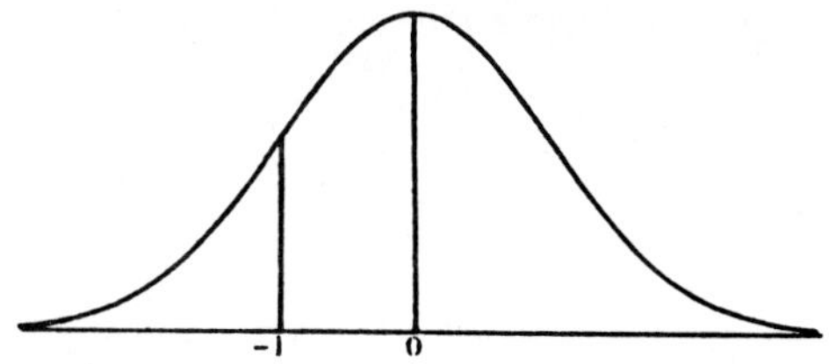

5-19 The probability of a reading less than 3.05 is the probability of a reading between 0 and 3.05 plus the probability of one less than 0. (See the figure below) From the table the probability of a reading between 0 and 3.05 is .4989. Then the probability of a reading less than 3.05 is .5 + .4989 = .9989.

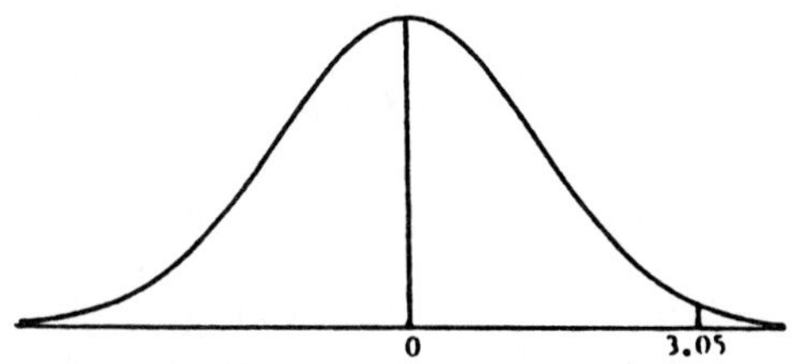

5-21 As we see in the figure below, these two are on opposite sides of 0. Therefore, their respective probabilities must be added. From the table the probability for -1. is .3413 and that for 2. is .4772. Then the probability of lying between the two is .3413 + .4772 = .8185.

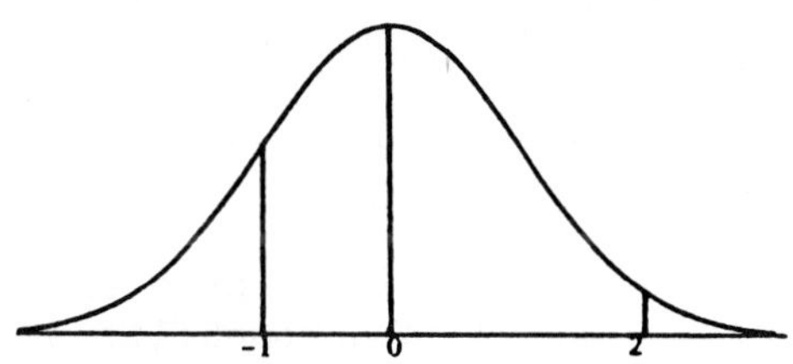

5-23 As in 5-21, we must add the probabilities for the two readings because they are on opposite sides of 0. The two probabilities are .4772 and .4332 which are .9104 when added.

5-25 As we see in the figure below, the two z values are on the same side of the mean. As a result we find the probability of lying between 0 and 2.0 and must subtract from that the probability of lying between 0 and 1. The two respective probabilities are .4772 and .3413. Their difference is .1359.

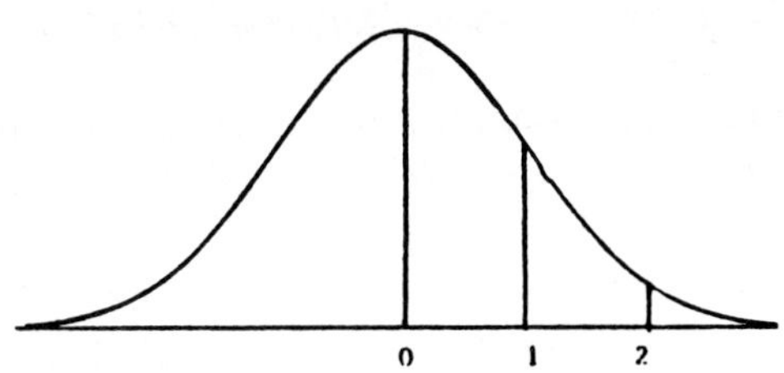

5-27 As in 5-25, these two readings are both on the same side of the mean. As a result, we find the respective probabilities for the z values and subtract them. The two respective probabilities are .4951 and .3997. The difference is .0954.

5-29 As we see in the figure below, the two z values are on the same negative side of the mean. Therefore, the probability of lying between the two readings is the probability of lying between 0 and -.83 minus the probability of lying between 0 and -.51. The respective probabilities are .2967 and .1950. The difference is .1017.

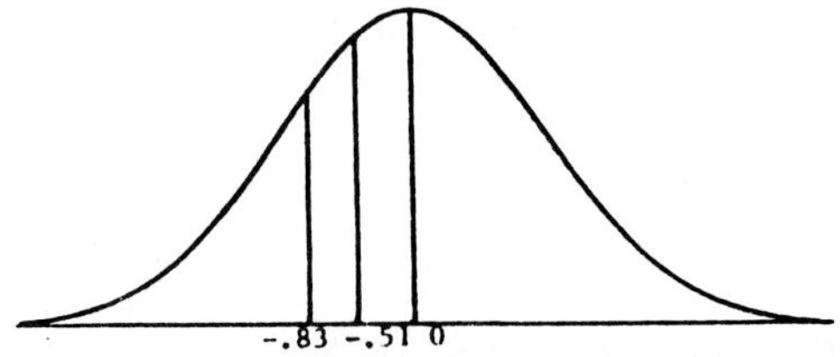

5-31 As in 5-29, both of these two are on the negative side of the mean. Therefore, we must look up their respective probabilities and subtract them. The probabilities are .4115 and .0987. The difference is .3128.

5-33 Since one half of the data lies above the mean, the probability is .5.

5-35 The probability of less than -.5 is .5 minus the probability of lying between -.5 and 0. The probability of more than 1.5 is .5 minus the probability of between 0. and 1.5. From the table these are .5 - .1915 = .3085 and .5 - .4332 = .0668 respectively. We must add the two probabilities to find the answer. This is .3085 + .0668 = .3753.(See the figure below)

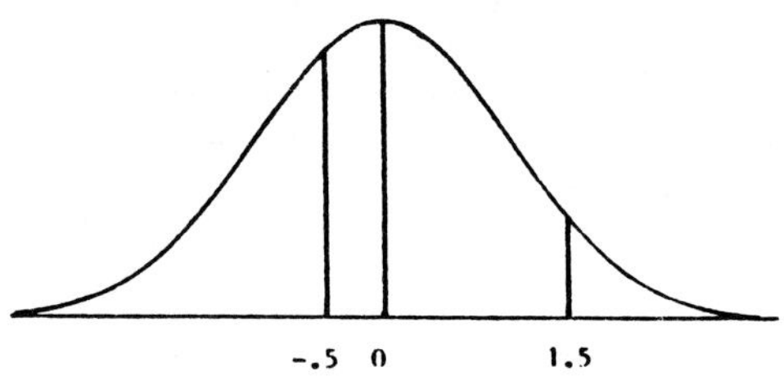

5-37 $P(z > 2.58)$ is the probability that z is more than 2.58. As we see in the figure, this is .5 minus the probability that z lies between 0 and 2.58 or $.5 - P(0 < z < 2.58) = .5 - .4951 = .0049$.

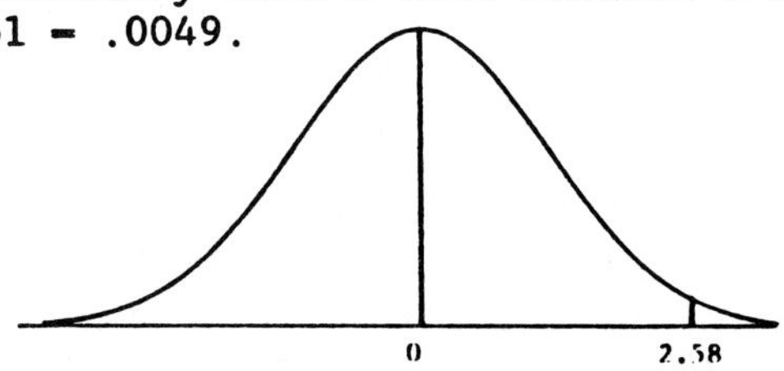

5-39 $P(z < 1.28)$ is the probability that z is less than 1.28. This is $.5 + P(0 < z < 1.28) = .5 + .3997 = .8997$.

5-41 $P(z < -0.57)$ is the probability that z is below -.57. As seen in the figure, this is $.5 - P(-0.57 < z < 0.) = .5 - .2157 = .2843$.

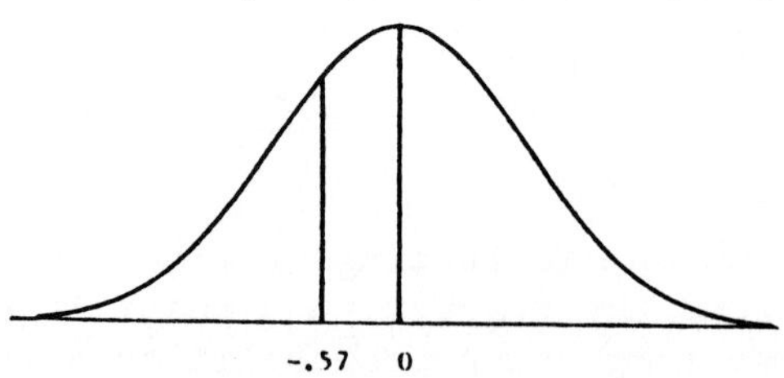

5-43 Since these are on the same side of the mean, their probabilities must be subtracted. That is,

$$P(-2.8 < z < -1.36) = P(-2.8 < z < 0.) - P(-1.36 < z < 0)$$

$= .4974 - .4131 = .0843.$

5-45 Using symmetry, we have directly from the table

$P(-1.09 < z < 0) = P(0 < z < 1.09) = .3621.$

5-47 As in 5-39,

$P(z < 2.45) = .5 + P(0 < z < 2.45) = .5 + .4929 = .9929.$

SOLUTIONS TO 5-1 B EXERCISES

5-49 a. This is $P(-1 < z < 1) = .3413 + .3413 = .6826$ since the z values are on opposite sides of the mean.

b. As in a. $P(-1.96 < z < 1.96) = .475 + .475 = .95.$

c. As in a. $P(-3 < z < 3) = .4987 + .4987 = .9974.$

d. $P(-1 < z < 2) - .3413 + .4772 = .8185.$

e. This is $P(z > 2) + P(z < -2)$. By symmetry the two are the same. $P(z > 2) = .5 - P(0 < z < 2) = .5 - .4772. = .0228$. We then double it. The answer is then .0456.

5-51 If the two tails are to be weighted equally, then there would be 1% on the upper tail and 1% on the lower. As seen in the figure, we would have 49% of the readings between 0 and the upper acceptable level. This corresponds to an upper level of 2.33. The lower level is by symmetry -2.33.

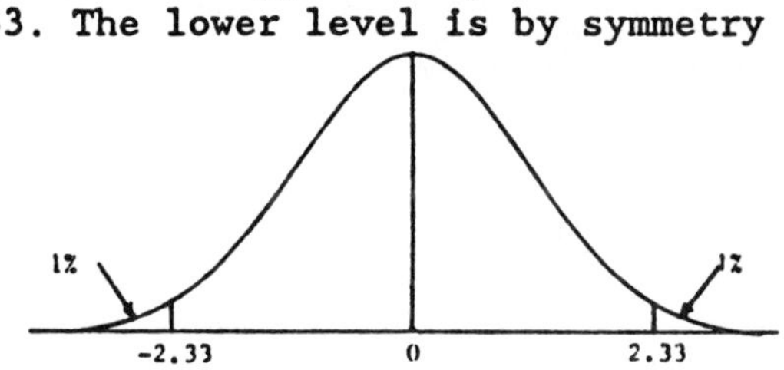

5-55 The probability of exceeding k + 1 in the translated distribution is the same as the probability of exceeding 1 in the standard normal distribution. This is $.5 - P(0 < z < 1) = .5 - .3413 = .1587.$

SOLUTIONS TO SECTION 5.3 - NONSTANDARD NORMAL DISTRIBUTION

5-57 Applying the formula to transform the data to a standard normal curve, we would have

$$z = \frac{x - \mu}{\sigma} = \frac{145 - 100}{15} = \frac{45}{15} = 3$$

and

$$z = \frac{x - \mu}{\sigma} = \frac{100 - 100}{15} = 0.$$

We are looking for the probability that z lies between 0 and 3.

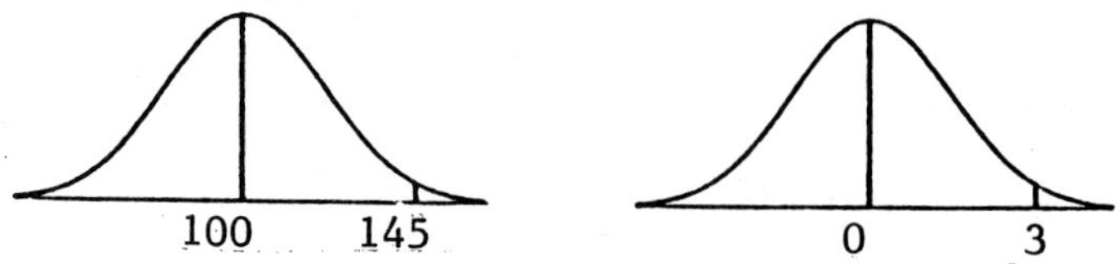

Using the normal table directly, this is .4987.

5-59 We want the probability that the random variable x exceeds 140. We compute z as

$$z = \frac{x - \mu}{\sigma} = \frac{140 - 100}{15} = \frac{40}{15} = 2.67.$$

Consequently we are looking for the probability that z is more than 2.67.

From the table we find that the probability of lying between 0 and 2.67 is .4962. The probability that z exceeds 140 is .5 - .4962 which is .0038.

5-61 To compute the probability of x lying between 110 and 120 we need the associated z values.

$$z = \frac{x - \mu}{\sigma} = \frac{110 - 100}{15} = .67.$$

$$z = \frac{x - \mu}{\sigma} = \frac{120 - 100}{15} = \frac{20}{15} = 1.33$$

The probability that z lies between these two is the probability that the value of z lies between 0 and 1.33 minus the probability that z lies between 0 and .67.

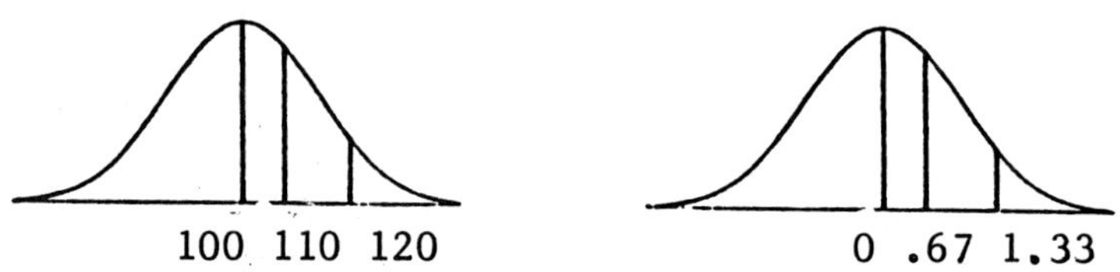

From the table of z scores this will be .4082 - .2486 = .1596.

5-63 The associated z value for x = 100 when z = 100 is z = 0. The probability that z is less than 0 is .5.

5-65 We are given that $\mu = 2.97$ and $\sigma = .60$. We are looking for the probability that x is between 2.97 and 3.87. These x values correspond to z values of 0 for 2.97 and of

$$z = \frac{x - \mu}{\sigma} = \frac{3.87 - 2.97}{.6} = \frac{.9}{.6} = 1.5$$

We want the probability that z is between 0 and 1.5. From the table this is .4332.

5-67 We want the probability that x lies between 60 and 72. As z scores we get

$$z = \frac{60 - 72}{15} = -.80 \quad \text{and} \quad z = \frac{72 - 72}{15} = 0.$$

Consequently we want the probability that z lies between .8 and 0. To find this it is the same as the probability that z lies between 0 and .8.

From the table this will be .2881.

5-69 a. We are looking for the probability that x exceeds 4. Because $\mu = 3.8$ and $\sigma = .95$, the probability that x is more that 4 is the probability that z is more than

$$z = \frac{x - \mu}{\sigma} = \frac{4. - 3.8}{.95} = \frac{.2}{.95} = .21$$

From the table this would correspond to a probabilitiy of .0832. of lying between 0 and .21. The final answer of exceeding .21 will be .5 - .0832 = .4168 or 41.68%.

5-71 The probability that a score exceeds 650 is the probability that z exceeds (650 - 500)/100 = 1.5. This is .5 minus the probability that z lies between 0 and 1.5. Then p = .5 - .4332. This will be .0668 or 6.68%.

5-73 Given $\mu = 178.1$ and $\sigma = 40.7$, we want the probability of a cholesterol level lying between 100 and 200. The associated z scores are

$$z = \frac{100 - 178.1}{40.7} = -1.92 \quad \text{and} \quad z = \frac{200 - 178.1}{40.7} = .54.$$

As seen in the figure below, since the z values are on opposite sides of thehe mean, we add the respective probabilities.

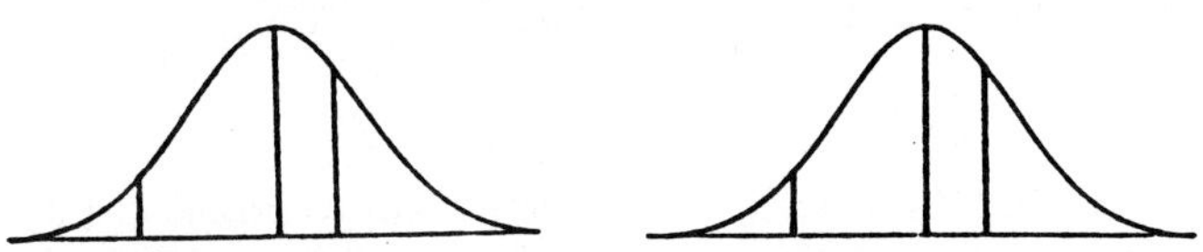

From the table the probabilities associated with the respective z values are .4726 and .2054 respectively. Consequently the probability is .4726 + .2054 = .6780.

5-75 For the x limits of 110 and 150, the associated z values are

$$z = \frac{x - \mu}{\sigma} = \frac{110 - 99.56}{25.84} = \frac{10.44}{25.84} = .40.$$

and

$$z = \frac{x - \mu}{\sigma} = \frac{150 - 99.56}{25.84} = \frac{50.44}{25.84} = 1.95.$$

Since these are on the same side of the mean as seen in the figure below, the two associated probabilities must be subtracted. Therefore, the probability is .4744 - .1554 = .3190.

5-77 This problem is solved in a manner similar to 5-73.

5-79 To compute the probability of not qualifying, we need the probability of a score below 172. We compute the associated z score. This is

$$z = \frac{x - \mu}{\sigma} = \frac{172 - 119.3}{32.4} = \frac{52.7}{32.4} = 1.63.$$

The probability that a score is below 172 is the probability that z is below 1.63 which is .5 + P(z is between 0 and 1.63) = .5 + .4484 = .9486.

5-81 Using $\mu = 61.3$ and $\sigma = 3.3$, we are trying to determine the probability that a speed lies between x values of 55.0 and 60.0. These correspond to z values of

$$z = \frac{55 - 61.3}{3.3} = -1.91 \text{ and } z = \frac{60 - 61.3}{3.3} = -.39$$

respectively. As we see in the figure below both z values are on the same side of the mean, as a result, we must subtract the two probabilities. From the table the answer will be .4719 - .1517 = .3202.

5-83 We know that $\mu = 39.4$ weeks and $\sigma = 2.43$ weeks. We are looking for the probability of being more than 3 weeks from the mean. This is 1 minus the probability of being within 3 weeks which is twice the probability of lying in the interval from 39.4 to 42.4. Then,

$$z = \frac{42.4 - 39.4}{2.43} = \frac{3}{2.43} = 1.23.$$

The associated probability is .3907. As a result the probability is .3907 + .3907 = .7814. The probability of being outside of 3 weeks is then 1 - .7814 = .2186.

5-85 The probability of exceeding 16 is the probability of a z value greater than

$$z \frac{16 - 18.4}{5.1} = -.47.$$

This is .5 plus the probability that z lies between -.47 and 0. From the table this latter is .1808. As a result the final probability is .6808. Then 68.08% of the 50 students is 34.04.

5-87 This exercise is solved like 5-69.

SOLUTIONS TO 5-3 B EXERCISES

5-89 a. We find the mean by summing the data values and dividing by the number of observations n = 30. The result is that the mean is 50.47.

b. The standard deviation is determined using the formula

$$s^2 = \frac{n(\sum x^2) - (\sum x)^2}{n(n-1)} = \frac{30(76544) - 2292196}{30(29)} = 4.74.$$

Then the standard deviation is 2.2.

c. Since there are 14 of the 30 scores in that range, the percentage is 46.67%.

d. Translating the x values of 51 and 54 into z scores, assuming a mean of 50.47 and a standard deviation of 2.2, we would have

$$z = \frac{51 - 50.47}{2.2} = .24 \quad \text{and} \quad z = \frac{54 - 50.47}{2.2} = 1.6.$$

Since these two z values are on the same side of the mean we would subtract the associated probabilities. The result is .4452 - .0948 = .3504 or 35.04%.

5-91 For the uniform distribution the range of interest is 80 to 110 with a total range of 74 to 126. Consequently the probability as determined by the ranges is 30/52 = .577.

For the normal distribution the two associated z values are

$$z = \frac{80 - 100}{15} = -1.33 \quad \text{and} \quad z = \frac{110 - 100}{15} = .67.$$

The associated probabilities are .4082 and .2486. Since the z values are on opposite sides of the mean, we determine the final probability as their sum or .4082 + .2486 = .6568.

SOLUTIONS TO SECTION 5-4 FINDING SCORES WHEN GIVEN PROBABILITIES

5-93 As we see in the picture, if 95% are below some value, then 45%

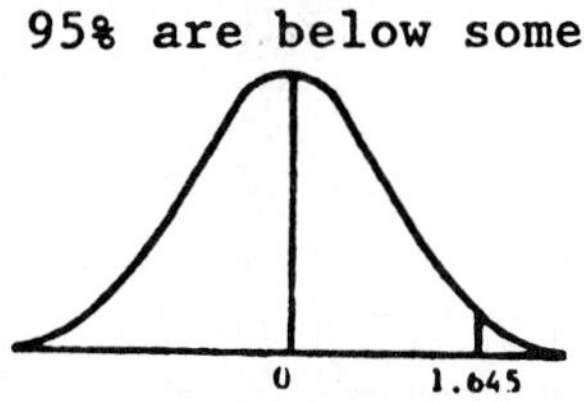

will lie between 0 and that value. Looking up .45 in the table, we will arrive at a z value of 1.645.

5-95 Because of symmetry, we refer to 5-93 above and conclude that the z value is -1.645.

5-97 The acceptable range would exclude the top 5% as was determined in 5-73 and the bottom 5% as was determined in 5-75. Consequently the acceptable range is from -1.645 to + 1.645.

5-99 As we see in the figure, to determine the level of the top 10% we

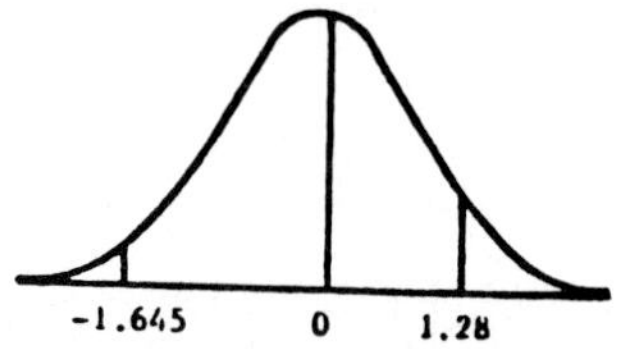

must use the fact that 40% lie between 0 and the associated z value. Looking up .40 in the normal table we will arrive at a z value of 1.28. The bottom 5% was determined in 5-95. The range of acceptable readings extends from -1.645 to 1.28.

5-101 As seen in the picture, we must look up a probability in the table of .10. This is associated with the z value of .25.

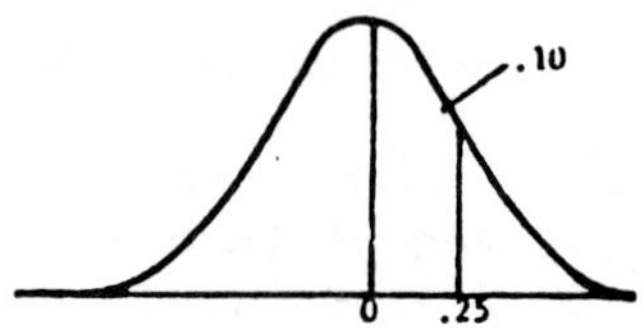

5-103 As seen in the picture, we must look up a probability in the table of .25. This is associated with a z value of .675.

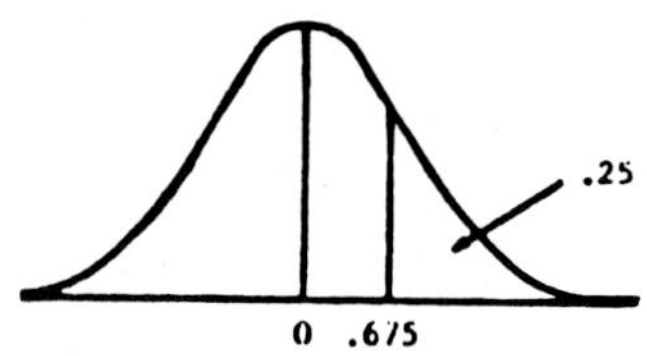

5-105 The z value associated with the upper 5% is 1.645 as was found in problem 5-93. To find the associated value, x, of measure of burnout, we form

$$z = \frac{x - \mu}{\sigma} \quad \text{or} \quad 1.645 = \frac{x - 2.97}{.6}.$$

Solving for x we have

$$x - 2.97 = (1.645)(.6) = .987,$$

or $x = 3.957$.

5-107 The z value associated with the the seventh decile is the z value associated with the probability of .20 above the mean, as seen in the figure.

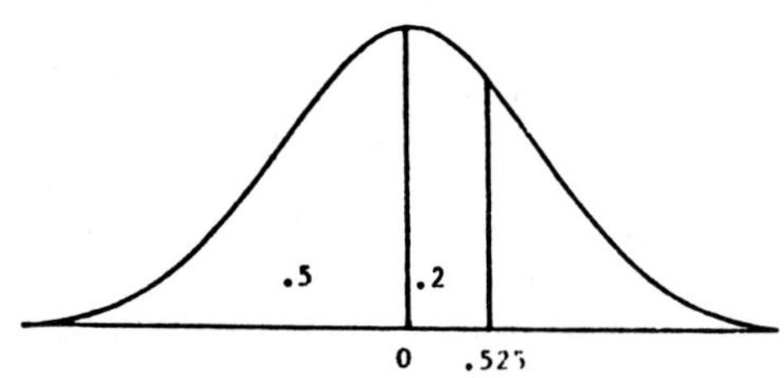

Then,

$$z = \frac{x - \mu}{\sigma} \quad \text{or} \quad .525 = \frac{x - 72}{15}.$$

Then $x - 72 = (15)(.525) = 7.875$, or $x = 79.875$.

5-109 We are looking for the height which separates the upper 95% from the lower 5%. This was determined in problem 5-95 to be -1.645. Using the formula of 5-107, the minimum height will be

$$x = \mu + z\sigma = 69 + (-1.645)(2.8) = 64.394.$$

5-111 This problem is solved like 5-107, except the top 67% level gives a z value of -.44. As a result the x value is

$$x = \mu + z\sigma = 4.7 + (-.44)(.8) = 4.348.$$

5-113 The top 5% and the bottom 5% are associated with z values of 1.645 and -1.645 respectively. These were determined in 5-97. Then the two associated x values are

$$x = \mu + z\sigma = 100 + (1.645)(15) = 124.675$$

and

$$x = 100 + (-1.645)(15) = 75.325.$$

5-115 This exercise is solved like 5-113, except that because of the deciles, we want the top 10% and the bottom 10%.

5-117 The z value associated with the top 7% comes from a table value of .43. The z value is 1.475. To determine the z value for the bottom 13% we would look for a table value of .37. The associated z value is -1.13 since it is a lower 13%. The associated x values are

$$x = \mu + z\sigma = 25.1 + (1.475)(.2) = 25.395.$$

and

$$x = 25.1 + (-1.13)(.2) = 24.874.$$

These are then the limits of acceptable ball bearing diameters.

5-119 The method of solution here is the same as in 5-117.

SOLUTIONS TO 5-4 B EXERCISES

5-121 We are trying to determine the fastest 200 times out of the 4830 applicants. This will amount to the top 4.14%. The probability to be used for the table is .4586. The associated z value is -1.735 because the fastest times are on the left tail of the curve. Then the time would be

$$t = \mu + z\sigma = 36.2 + (-1.735)(3.8) = 29.6 \text{ minutes.}$$

5-123 a. Using the formulas of chapter 2 we have a sample mean of

$$\bar{x} = \frac{\sum x}{n} = \frac{2723}{40} = 68.075.$$

b. For the standard deviation we compute the variance as

$$s^2 = \frac{n\sum x^2 - (\sum x)^2}{n(n-1)} = \frac{40(189263) - (2723)^2}{(40)(39)} = 99.87.$$

As a result the standard deviation is $s = 9.99$.

c. The top 5% of 40 scores will be the top 2 scores. The lowest of these is 82.

d. Using the sample statistics as estimates of μ and σ we will have a score level of

$$x = 68.075 + (1.645)(9.99) = 84.51$$

where we used a z value associated with the upper 5% of 1.645.

SOLUTIONS TO SECTION 5-5 NORMAL AS APPROXIMATION TO BINOMIAL

5-125 We have $np = 25 \times .25 = 6.25$ and $nq = 25 \times .75 = 18.75$. Since both are more than 5, the normal distribution is a suitable approximation. We can compute $\mu = np = 6.25$ and $\sigma = \sqrt{npq}$. To compute this, we have

$$\sigma = \sqrt{(25)(.25)(.75)} = 2.165.$$

5-127 We have np = (90)(.048) = 4.32. Since this is less than 5, the normal distribution is not a suitable approximation to the binomial. We have $\sigma =$ 2.03.

5-129 We have np = (250)(.010) = 2.5 and nq = (250)(.99) = 225. Since one of these does not exceed 5, the normal distribution will not be a good approximation to the binomial. We compute

$\mu = 2.5$ and $\sigma = \sqrt{npq}$ which is $\sigma = \sqrt{(250)(.01)(.99)} = 1.57$.

5-131 We have np = (84)(.95) = 79.8 and nq = (84)(.05) = 4.2. Since nq is less than 5, the normal distribution will not be a good approximation to the binomial.

5-133 a. Using the table the probability is .121.

b. Using the normal distribution we want the probability of 8, which with the continuity correction is the probability that the normal random variable lies between 7.5 and 8.5.We have
$\mu = np = (12)(.5) = 6$ and $\sigma = \sqrt{(12)(.5)(.5)} = 1.732$.
The z values associated with x values of 7.5 and 8.5 are

$$z = \frac{x - \mu}{\sigma} = \frac{7.5 - 6}{1.732} = .87 \quad \text{and} \quad z = \frac{8.5 - 6}{1.732} = 1.44$$

respectively. We look the two z values up in the normal table and find probabilities of .3078 and .4251. Both of these two z values are to the right of the mean, and so the probability of lying between them is the difference of the two probabilities. Consequently, the normal approximation to the correct binomial probability is .4251 - .3078 = .1173.

5-135 a. From them the binomial table the correct probability is .114.

b. We are seeking the probability of 12. Then the normal approximation would determine the probability that the normal random variable x will lie in the interval from 11.5 to 12.5.

We have $\mu = np = (20)(.7) = 14.$ and $\sigma = \sqrt{(20)(.7)(.3)} = 2.05$.
Then the two z values are

$$z = \frac{11.5 - 14}{2.05} = -1.22 \quad \text{and} \quad z = \frac{12.5 - 14}{2.05} = -.73.$$

When we look these two z values up in the normal table we find probabilities of .3888 and .2673 respectively. Since we want the probability of lying between these two z values and they are both on the same side of the mean, we must subtract the two probabilities. Then the final approximating probability is .3888 - .2673 = .1215.

5-137 a. The probability of at least 8 from the table is the sum of the probabilities of 8, 9, 10, 11, and 12. Adding all of these we find a binomial probability of .194.

b. The probability of at least 8 is the probability of 8 or more. Since this includes 8, we are looking for the normal approximating probability of x greater than or equal to 7.5. We compute the mean and standard

deviation as
$\mu = np = (12)(.5) = 6$ and $\sigma = \sqrt{npq} = \sqrt{(12)(.5)(.5)} = 1.732$.
Then $z = \frac{x - \mu}{\sigma} = \frac{7.5 - 6}{1.732} = .87$. Since this z value exceeds

the mean, then the probability that the random variable is more than this is .5 minus the probability that it lies in the interval between 0 and .87. Therefore the estimated probability is given by .5 - .3078 = .1922.

5-139 a. The probability of at most 12 is the probability that we found 12 or fewer. Adding the probabilities from the table yields a probability of .227.

b. When we determine the region for the normal approximation, we include 12, consequently we wish to determine the probability that the random variable lies in the interval less than 12.5. We compute the normal random variable as

$$z = \frac{x - \mu}{\sigma} = \frac{12.5 - (20)(.7)}{\sqrt{(20)(.7)(.3)}} = -.73.$$

We are determining the probability that z will be less than -.73, which is .5 - .2673 = .2327.

5-141 The probability of a birth being a boy is .5. Then np = 50 and nq = 50. Since both exceed 5, then the normal approximation is appropriate. The probability of at least 60 is the probability of 60 or more, which includes 60. Therefore we are to find the probability that the normal random is 59.5 or more.

$\mu = np = 50$ and $\sigma = \sqrt{npq} = \sqrt{(100)(.5)(.5)} = 5$
and

$z = \frac{59.5 - 50}{5} = 1.9$. The probability that z exceeds 1.9 is .5 minus the probability that z is in the interval from 0 to 1.9.

From the normal table this is .5 - .4713 = .0287.

5-143 Since p = .5 of guessing correctly, then np = 50 and nq = 50. Both of these exceed 5, therefore we can use a normal approximation to the binomial. The probability of at least 65 correct answers includes 65, consequently we want the probability that the normal random variable is more than 64.5. Computing the normal variable z we have

$$z = \frac{64.5 - 50}{\sqrt{(100)(.5)(.5)}} = \frac{14.5}{5} = 2.9.$$

We are looking for the probability that z is more than 2.9. This will be .5 - .4981 = .0019.

5-145 Since p = .72, then np = 57.6 and nq = 12.96. Both of these exceed 5 so a normal approximation is appropriate. The probability of fewer than 70 is the probability that x is less than 69.5. Computing z we get

$$z = \frac{69.5 - 57.6}{\sqrt{(80)(.72)(.18)}} = \frac{11.9}{3.22} = 2.96.$$

The probability of a z value less than 2.96 is .5 plus the probability that z lies between 0 and 2.96. The resultant probability is .5 + .4985 =

.9985.

5-147 Solved like 5-145 using $p = .25$, $n = 400$, and a normal variable x of 124.5 or more.

5-149 Since the probability is .25 of kickback type accident, then the probability of another type is .75. Also $n = 84$. Since $np = 21$ and $nq = 63$, both of which exceed 5, the normal approximation is appropriate. We are seeking the probability that x equals 21. That is the probability that the normal variable is between 20.5 and 21.5. Computing z we have

$$z = \frac{20.5 - 21}{3.97} = -.13 \quad \text{and } z = \frac{21.5 - 21}{3.97} = .13.$$

Since these are on opposite sides of 0, then we must add their respective probabilities. Then the probability is .0517 + .0517 = .1034.

5-151 We have $p = .60$, $n = 500$ and $np = 300$. We are looking for the probability of more than 325 men living at home. This is approximated by the probability that the normal random variable, x, is more than 325.5. In terms of the standard normal curve this is the probability that z is more than

$$\frac{x - \mu}{\sigma} = \frac{x - np}{\sqrt{npq}} = \frac{325.5 - 300}{10.95} = 2.33.$$

The probability associated with a z value of 2.33 is .4901. This is subtracted from .5 to obtain the probability that z exceeds the given level. The final probability is .5 - .4901 = .0099.

5-153 With $p = .75$, $n = 320$, then $np = 240$. We want the probability that the normally distributed random variable is more than 250.5. From here the problem proceeds like 3-151.

5-155 Using $p = .25$, and $n = 40$, we would be seeking the probability that the normal variable is less than 4.5. The solution from this data proceeds like 5-151.

5-157 With $p = .065$ and $n = 175$, the the mean is $np = 11.375$. We want the probability that the binomial x lies between 10 and 15 inclusive. This is the probability that the normal approximation lies in the interval from 9.5 to 15.5 since it is inclusive. We get the z values as

$$z = \frac{x - \mu}{\sigma} = \frac{9.5 - 11.375}{3.26} = -.58$$

and

$$z = \frac{x - \mu}{\sigma} = \frac{15.5 - 11.375}{3.26} = 1.26.$$

Since these are on opposite sides of the mean, their respective probabilities must be added. From the table the answer is the sum of .2190 + .3962 = .6152.

5-159 This exercise proceeds in the same manner as 5-157.

SOLUTIONS TO 5-5 B EXERCISES

5-161 a. From the table the probability is .001.

b. From the formula we have

$$P(18) = \frac{25!}{18!7!}(.4)^{18}(.6)^{7}$$

To this we must add P(19)+P(20)+P(21)+P(22)+P(23)+P(24)+P(25) all computed using the binomial formula. The final probability is .0012.

5-163 We are interested in a certain number of people showing. Then p(show) = .93. We do not know n = # of tickets sold in this problem. We do know n is at least 250. Thus np is at least (250)(.93) = 232.5 and nq is at least (250)(.07) = 17.5, and so we can use normal curve methods. We are looking for P(if show will have seat) is at least .95. Thus we want the variable x to be the number who show out of n = number of tickets sold, $P(x \le 250)$ is at least .95. For the normal curve approximation including 250 seats we want $P(x \le 250.5)$ is at least .95. Since $\mu = .93n$ and $\sigma = \sqrt{(.07)(.93)n} = .274\sqrt{n}$. We want

$$P\left(z \le \frac{250.5 - .93n}{.274\sqrt{n}}\right) \le .95.$$

Then for z_0 such that $P(z \le z_0) \le .95$ would mean

$z_0 \le 1.64$ from the normal table. Thus

$$\frac{250.5 - .93n}{.274\sqrt{n}} \le 1.64$$

$$250.5 - .93n \le .449\sqrt{n}$$

or

$$.93n + .449\sqrt{n} - 250.5 \le 0.$$

This is a quadratic equation in $\sqrt{n}$ and is solved using the quadratic equation. From that

$$\sqrt{n} = \frac{-.449 + \sqrt{(.449)^2 - (4)(.93)(-250.5)}}{(2)(.93)}$$

$$= \frac{-.449 + \sqrt{.20 + 931.86}}{1.86}$$

$$= \frac{-.449 + 30.53}{1.86} = \frac{30.08}{1.86} = 16.17.$$

Thus $n = (16.2)^2 = 262.5$. Since 263 would be too large, we round down to $n = 262$.

SOLUTIONS TO SECTION 5-6 THE CENTRAL LIMIT THEOREM

5-165 a. To find the probability of a single sample we need the associated z value.

$$z = \frac{x - \mu}{\sigma} = \frac{53 - 50}{10} = .3$$

The other z value is at the mean and so it is 0. Therefore we need the probability of a z value between 0 and .3 which from the normal table is .1179.

b. With n = 36, then

$\sigma_{\bar{x}} = \frac{\sigma}{\sqrt{n}} = \frac{10}{6} = 1.67$. To find the z values associated with the two levels of x, we note that one is 50, which is the mean. As a result, the z value is 0. For the other we have

$$z = \frac{\bar{x} - \mu}{\sigma_{\bar{x}}} = \frac{53 - 50}{1.67} = 1.80.$$

Then we want the probability that z lies between 0 and 1.80. From the normal table this is .4641.

5-167 a. We find the probability of the event by determining the z value.

$$z = \frac{x - \mu}{\sigma} = \frac{4.5 - 5.}{1.05} = .48$$

Since we want the probability that z is less than .48, we add .5 to the probability that z is between 0 and .48 or .5 + .1844 = .6844.

b. For this we must compute the standard deviation of the mean as.

$\sigma_{\bar{x}} = \frac{\sigma}{\sqrt{n}} = \frac{1.05}{6.32} = .17$. Then $z = \frac{\bar{x} - \mu}{\sigma_{\bar{x}}} = \frac{4.5 - 5}{.17} = 2.94$.

Therefore we are looking for the probability that z lies between 0 and 2.94. From the normal table this is .4984 which must be added to .5 as in a. above. The final probability is .9984.

5-169 This exercise is solved like 5-165 using $n = 36$, $\mu = 100$ and $\sigma = 15$.

5-171 We have

$\mu = 20$ and $\sigma_{\bar{x}} = \frac{\sigma}{\sqrt{n}} = \frac{20}{6} = 3.33$. As a result z is given by

$$z = \frac{\bar{x} - \mu}{\sigma_{\bar{x}}} = \frac{275 - 280}{3.33} = -1.5.$$

The probability that x exceeds 50 is the probability that z exceeds -1.5. This is .5 + .4332 = .9332.

5-173 The population is assumed to be large so the normal distribution can be used. We have

$\mu = 10.7$ and $\sigma_{\bar{x}} = \frac{11.2}{6.48} = 1.73$. The value of z is

$$z = \frac{12 - 10.7}{1.73} = .75 .$$

The probability that the average is less than 12 is .5 + the probability that z is between 0 and .75. This is .5 + .2734 = .7734.

5-175 The sample size of 33 is more than 5 and less than 5% of the population size of 800. As a result we use the normal distribution to give approximate probabilities. Computing z we have

$\mu = .82$ and $\sigma_{\bar{x}} = \frac{\sigma}{\sqrt{n}} = \frac{.18}{5.74} = .031$. As a result z is given by

$$z = \frac{.825 - .82}{.031} = .16.$$

The probability that the sample average is less than .825 is .5 + the probability that z lies between 0 and .16. From the table this is .5 + .0636 = .5636.

5-177 This exercise is solved in a manner simila to 5-175.

5-179 The sample of 31 is more than 5 and less than 5% of the population. This meets the criteria for us to use normal approximations. The two z values associated with the two means are

$$z = \frac{\bar{x} - \mu}{\sigma_{\bar{x}}} = \frac{3.0 - 2.97}{.18} = .28$$

and

$$z = \frac{3.1 - 2.97}{.18} = 1.20.$$

We subtract the two associated probabilities to obtain the final probability because they are on the same side of the mean. The two probabilities are .1107 and .3849. The final probability is then .3849 - .1107 = .2742.

5-181 We compute the standard errors for the two sample sizes.

$$\sigma_{\bar{x}} = \frac{\sigma}{\sqrt{n}} = \frac{20}{10} = 2 \quad \text{and} \quad \sigma_{\bar{x}} = \frac{20}{20} = 1.$$

As a result an increase of the sample by a factor of 4 gave a decrease in the standard error by a factor of 2.

5-183 a. 5% of N = 5000 is 250. Therefore we do not need to use the correction factor.

$$\sigma_{\bar{x}} = \frac{\sigma}{\sqrt{n}} = \frac{15}{\sqrt{200}} = 1.06.$$

b. 5% of 12000 is 600. Since n = 1000 exceeds this we do need to use the correction factor of

$$\sqrt{\frac{12000 - 1000}{12000 - 1}} = \sqrt{.9167} = .957.$$

Then

$$\sigma_{\bar{x}} = \frac{.957\sigma}{\sqrt{n}} = .454.$$

c. 5% of 4000 is 200. Since n = 500 exceeds this, we must use the correction factor of

$$\sqrt{\frac{4000 - 500}{4000 - 1}} = \sqrt{.8752} = .9355.$$

$$\sigma_{\bar{x}} = \frac{.9355\sigma}{\sqrt{n}} = .628.$$

5-185 The sample size of 36 is more than 5% of the population size of 160. As a consequence the correction factor must be used. It is

$$\sqrt{\frac{160 - 36}{160 - 1}} = .78.$$

The associated z values are

$$z = \frac{\bar{x} - \mu}{\sigma_{\bar{x}}\,(.78)} = \frac{7. - 8.5}{.66(.78)} = -2.91.$$

and

$$z = \frac{\bar{x} - \mu}{\sigma_{\bar{x}}\,(.78)} = \frac{10. - 8.5}{.66(.78)} = 2.91.$$

The probability that the average is between 7 and 10 is same as the probability that z is between -2.91 and 2.91. The associated probabilities must be added because the values are on opposite sides of the mean. Thus the final answer is .4982 + .4982 = .9964.

5-187 The selection of 20% of the population gives a sample size of 50. We must use the correction factor here of

$$\sqrt{\frac{250 - 50}{250 - 1}} = .896.$$

The associated z values are

$$z = \frac{430 - 436}{3.39(.896)} = -1.98, \quad \text{and } z = \frac{440 - 436}{3.39(.896)} = 1.31.$$

Then the probability that the mean lies in the interval from 430 to 440 is the same as the probability that z lies in the interval from -1.98 to 1.31, which is .4761 + .4049 = .8810.

SOLUTIONS TO 5-6 B EXERCISES

5-189 a. Using the formula, the mean of this population is

$$\mu = \frac{\sum x}{n} = \frac{48}{6} = 8. \qquad \text{and } \sigma^2 = \frac{\sum(x-\mu)^2}{n} = \frac{174}{6} = 29. \text{ Therefore,}$$

$\sigma = 5.38$.

b. The samples will be - 2,3 2,6 2,8 2,11 2,18 3,6 3,8 3,11 3,18 6,8 6,11 6,18 8,11 8,18 11,18.

c. The means of the respective samples are 2.5, 4, 5, 6.5, 10, 4.5, 5.5, 7, 10.5, 7, 8.5, 12, 9.5, 13, 14.5.

d. Using the usual formula for the mean,

$$\mu = \frac{\Sigma\bar{x}}{n} = \frac{120}{15} = 8. \quad \text{Also, } \sigma^2_{\bar{x}} = \frac{\Sigma(\bar{x}-\mu)^2}{n} = \frac{174}{15} = 11.6. \text{ Therefore,}$$

$\sigma = 3.4$.

e. The two means are 8, and working with variances and using the sample size of n = 2,

$$\sigma^2_{\bar{x}} = \frac{174}{15} \text{ and } \frac{\sigma^2}{n}\frac{(N-n)}{N-1} = \frac{(174/6)}{2}\frac{(6-2)}{6-1} = \frac{174(4)}{2(5)(6)} = \frac{174}{15} = 11.6.$$

SOLUTIONS TO REVIEW EXERCISES

5-193 a. We compute the z value as

$$z = \frac{x - \mu}{\sigma} = \frac{5. - 5.4}{1.2} = -.33.$$

We look this up in the normal table and obtain a probability of .1293. To find the probability of exceeding this, we take .5+.1293 which is .6293.

b. Because the sample of 32 is less than 5% of the population, we do not need the correction factor. Therefore, the variance is

$$\sigma^2_{\bar{x}} = \frac{\sigma^2}{n} = \frac{1.44}{32} = .045. \text{ Then } z = \frac{\bar{x} - \mu}{\sigma} = \frac{5.-5.4}{.21} = -1.90.$$

Looking this up in the normal table, we get .4713. The probability is computed as .5 + .4713 = .9713.

c. We compute z by using the correction since the sample is 5% of the population.

$$\sigma_{\bar{x}} = \frac{1.2}{\sqrt{32}}\sqrt{\frac{700-35}{700-1}} = .202. \text{ Then } z = \frac{\bar{x} - \mu}{\sigma_{\bar{x}}} = \frac{5. - 5.4}{.202} = -1.98.$$

The probability that the mean is above 5.4 is the same as the probability that z is above -1.98. This is .5 + .4761 = .9761.

5-195 a. This comes directly from the table as .4222.

b. As we see in the figure, the probability of greater than -1.05 is the probability of lying between 0 and -1.05 + the probability of more than 0. This is .3531 + .5 = .8531.

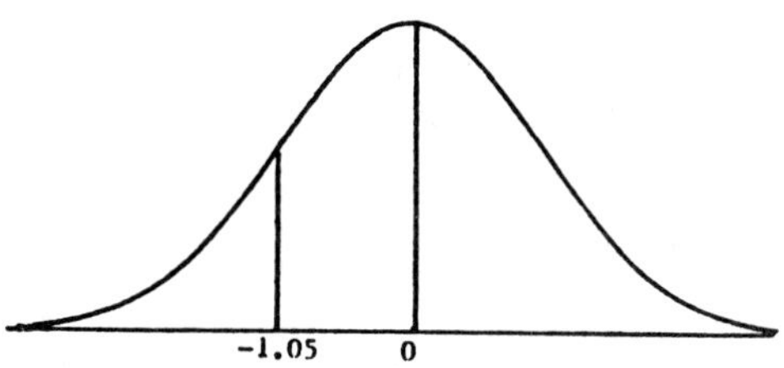

c. Since both of these are positive, the respective probabilities must be subtracted as seen in the figure. The final probability is .4332 - .1915 = .2417.

5-197 a. We compute the associated z as

$$z = \frac{x - \mu}{\sigma} = \frac{13.5 - 13.2}{2.95} = .10.$$

The probability of a z between 0 and .10 is .0398.

b. The probability of at least 12 years of education is the probability of a z value of more than -.41. Using the table this is .5 + .1591 = .6591.

c. The lowest quartile leads to a z value of -.675. The associated number of years of education will be $\mu + z\sigma$ = 13.2 + (.675)(2.95) = 11.21.

d. The top 5% is associated with a z value of 1.6456. Then the education level will be $\mu + z\sigma$ = 13.2 + (1.645)(2.95) = 18.05.

e. We compute z using

$$\sigma_{\bar{x}} = \frac{\sigma}{\sqrt{n}} = \frac{2.95}{5.92} = .498. \text{ Then } z = \frac{12 - 13.2}{.498} = -2.41.$$

Then the probability that the mean is 12 or more is the probability that z is more than -2.41. This is .5 + .4920 = .9920.

5-199 We use the normal approximation to the binomial here. The probability of at least 5 failures to respond is the probability that the normal variable, x, is at least 4.5. Using n = 150 and p =.04, we have μ = np = 6 and σ = 2.4. Then

$$z = \frac{4.5 - 6}{2.4} = -.63.$$

The probability that x is more than 5 is the probability that z is more than -.63. This is .5 + .2357 = .7357.

5-201 a. The two associated z scores are

$$z = \frac{x - \mu}{\sigma} = \frac{90 - 100}{15} = -.67 \quad \text{and} \quad z = \frac{x - \mu}{\sigma} = \frac{120 - 100}{15} = 1.33.$$

Since these are on opposite sides of the mean, the associated probabilities are added. Then, the probability is given by .2486 + .4082 = .6568.

b. We want P(x > 105). For this the z score is

$$z = \frac{x - \mu}{\sigma} = \frac{105 - 100}{15} = .33.$$

Then P(x > 105) = P(z > .33) = .5 - .1293 = .3707.

c. We want $P(\bar{x} > 105)$

$$\sigma_{\bar{x}} = \frac{\sigma}{\sqrt{n}} = \frac{15}{5.48} = 2.74 \quad \text{and} \quad z = \frac{\bar{x} - \mu}{\sigma_{\bar{x}}} = \frac{105 - 100}{2.74} = 1.82$$

Then $P(\bar{x} > 105)$ = P(z > 1.82) = .5 - .4656 = .0344.

d. For this percentile we need the z value of a probability of .45. From the table this is 1.645. Then we solve for x in

$$z = \frac{x - \mu}{\sigma} \quad \text{or} \quad 1.645 = \frac{x - 100}{15}.$$

Then x = 100 + (15)(1.645) = 124.675.

e. We proceed as in d with a probability of .35 wielding a z value of -1.03.

5-203 We can use the normal approximation to the binomial probabilities of this problem. We have n = 100 and p = .32. The probability of 40 or fewer is the probability that the normal variable is less than 40.5. We compute μ = np = 32 and σ = 4.66. Then

$$z = \frac{40.5 - 32}{4.66} = 1.82.$$

The probability that x is less than 40.5 is the probability that z is less than 1.82. This is .5 + .4656 = .9656.

CHAPTER 6. TEST OF HYPOTHESIS

SOLUTIONS TO SECTION 6.1 - TESTING A CLAIM ABOUT A MEAN

6-1 a. The claim is that μ is more than 30 ($\mu > 30$). The alternative is $\mu \leq 30$. Since equality is with the latter, we must have

$H_0 : \mu \leq 30$

$H_1 : \mu > 30$

b. Similar to a).

c. The claim is that μ is at least 100 ($\mu \geq 100$). The alternative is $\mu < 100$. The equal sign is with the claim, so we have

$H_0 : \mu \geq 100$

$H_1 : \mu < 100$

d. Similar to c).

e. The claim is the mean μ is \$3271. ($\mu = 3271$). The alternative is $\mu \neq 3271$. Hence, using the equal sign with H_0, we have

$H_0 : \mu = 3271$

$H_1 : \mu \neq 3271$

6-3 a. The type I error is to accept that the mean is more than 30, when it is 30 or less. The type II error is to accept that the mean is 30 years or less when it is in fact more than 30.

b. Similar to a.

c. The type I error is to accept that the mean is under 100 when it in 100 or more. The type II error is to accept the mean as 100 or more when it is in fact less than 100.

d. Similar to c.

e. The type I error is to accept that the mean cost is not \$3271 when it is equal. The type II error is to accept a mean cost of \$3271 when it is not.

6-5 a. Since H_1 is $\mu > 30$, it is a right-tailed test.

b. Same as a).

c. Since H_1 is $\mu < 100$, it is a left-tailed test.

d. Same as c.

e. Since H_1 is $\mu \neq 3271$, it is a two-tailed test.

6-7 a. For a right tailed test with $\alpha = .05$, the critical value is 1.645.(See the figure below)

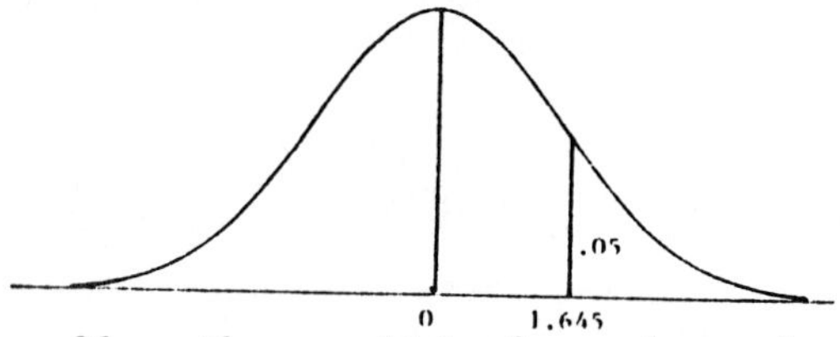

b. For a right tailed test with $\alpha = .01$, the critical value is 2.33. The figure is similar to a. above.

c. For a two-tailed test with $\alpha = .05$, the critical values are ± 1.96. The figure is below.

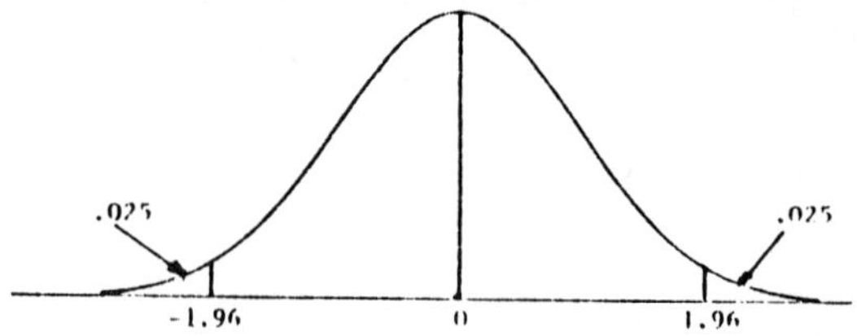

d. For a two-tailed test with $\alpha = .01$, the critical values are ± 2.575. The figure is similar to c.

e. For a left tailed test with $\alpha = .05$, the critical value is -1.645. The figure is below.

6-9 Following the outline from Figure 6-2, we are testing $\mu \leq 100$ with alternative $\mu > 100$. Since the equal sign is with $\mu \leq 100$, our test format is

$$H_0 : \mu \leq 100$$

$$H_1 : \mu > 100.$$

We are given $\alpha = .01$. We test the sample mean by computing

$$z = \frac{\bar{x} - \mu}{\sigma/\sqrt{n}} = \frac{100.8 - 100}{5/\sqrt{81}} = 1.44.$$

It is a right-tailed test with $\alpha = .01$ and the critical value is 2.33. Since $z = 1.44 < 2.33$, we conclude that the evidence does not support rejection of H_0 that $\mu \leq 100$.

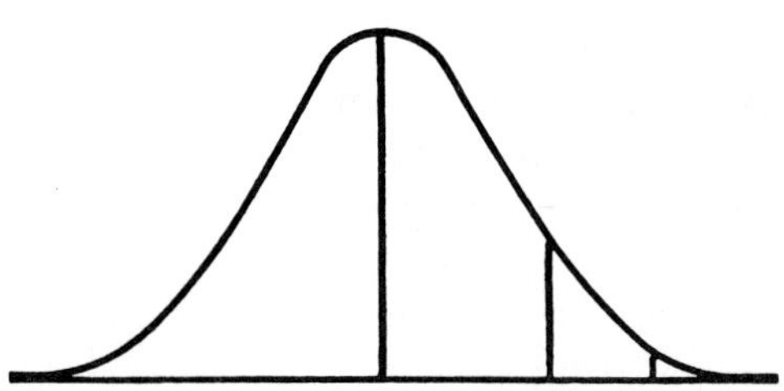

6-11 We are testing $\mu \geq 20$. The alternative is $\mu < 20$, so the form of the test is

$$H_0 : \mu \geq 20$$

$$H_1 : \mu < 20 \qquad \text{with } \alpha = .05.$$

We test the sample mean $\bar{x}$ by computing

$$z = \frac{\bar{x} - \mu}{\sigma/\sqrt{n}} = \frac{18.7 - 20}{3/\sqrt{100}} = -4.33.$$

It is a left-tailed test with $\alpha = .05$. The critical value is -1.645. Since $z = -4.33 < -1.645$ and thus in the critical region, we reject H_0 and conclude that there is sufficient evidence to support the claim that $\mu < 20$.

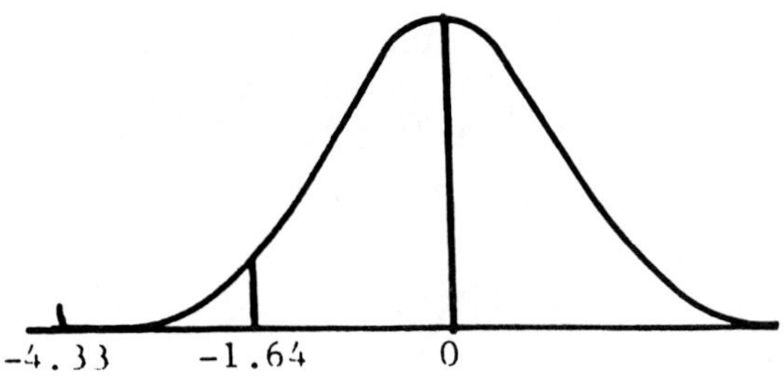

6-13 Our test is $\mu = 500$ and the alternative is $\mu \neq 500$. The test is given by

$$H_0 : \mu = 500$$

$$H_1 : \mu \neq 500 \qquad \alpha = .1.$$

Testing the sample mean we have

$$z = \frac{\bar{x} - \mu}{\sigma/\sqrt{n}} = \frac{510 - 500}{100/\sqrt{300}} = \frac{10}{5.77} = 1.73.$$

The test is a two-tailed test with $\alpha = .10$. The critical region is to the right of +1.645 or to the left of -1.645. Since z is in the critical region, we reject H_0. The evidence does not support the claim that $\mu = 500$.

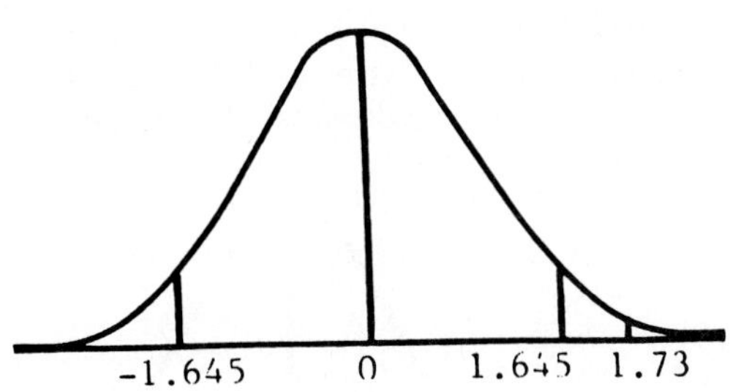

6-15 The claim is $\mu > 40$ and the alternative is $\mu \leq 40$. Hence we have

$H_0 : \mu \leq 40$

$H_1 : \mu > 40 \qquad \alpha = .05.$

Testing by using the sample mean $\bar{x}$, we have

$$z = \frac{\bar{x} - \mu}{\sigma/\sqrt{n}} = \frac{42 - 40}{8/\sqrt{50}} = \frac{2}{1.131} = 1.768.$$

This is a right-tailed test with $\alpha = .05$. The critical region is to the right of 1.645. Since 1.768 > 1.645, it is in the critical region and we reject H_0. We conclude that the evidence does not support the hypothesis of $\mu \leq 40$.

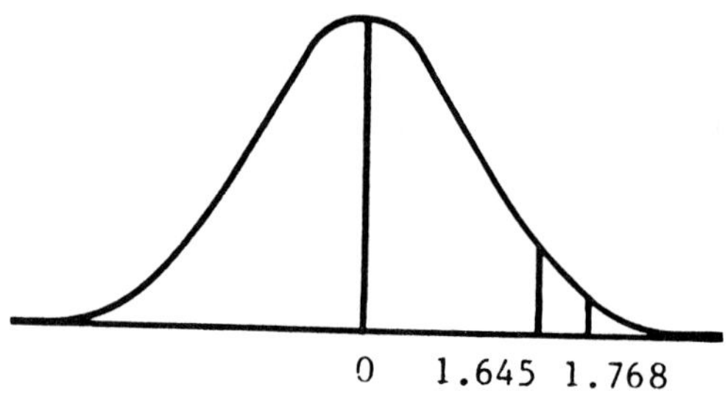

6-17 The claim is that the mean is more than 90000 ($\mu > 90000$). The alternative is $\mu \leq 90000$. The test is

$H_0 : \mu \leq 90000$

$H_1 : \mu > 90000 \qquad \alpha = .05.$

Using the sample average $\bar{x}$ we have

$$z = \frac{\bar{x} - \mu}{\sigma/\sqrt{n}} = \frac{96700 - 90000}{37500/\sqrt{191}} = \frac{6700}{2713.4} = 2.47.$$

This is a right-tailed test with $\alpha = .05$. The critical region is $z > 1.645$. Since our z value of 2.47 is in the critical region, we reject H_0. The evidence does support the claim that the mean is more.

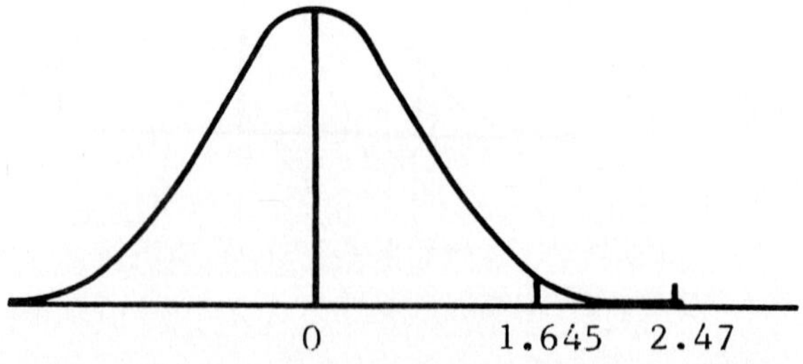

6-19 Our test is that $\mu < 7.5$ with alternative that $\mu \geq 7.5$. Then we have

$H_0 : \mu \geq 7.5$

$H_1 : \mu < 7.5 \qquad \alpha = .01.$

Computing our z score we have

$$z = \frac{\bar{x} - \mu}{\sigma/\sqrt{n}} = \frac{7.01 - 7.5}{3.74/\sqrt{100}} = \frac{-.49}{.374} = -1.31.$$

This is a left-tailed test with $\alpha = .01$. The critical region is less than -2.33. Since -1.31 is not in the critical region, we fail to reject H_0. The evidence does support the claim of a mean of 7.5 or more

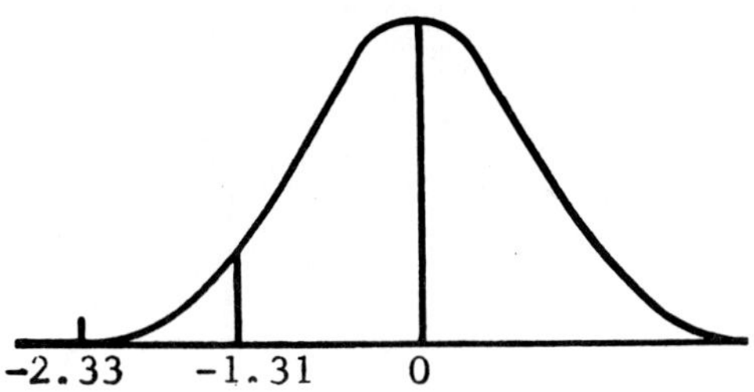

6-21 That the mean of the two regions is the same would be a test that $\mu = 13.2$. The alternative is $\mu \neq 13.2$. The test is then given by

$H_0 : \mu = 13.2$

$H_1 : \mu \neq 13.2 \qquad \alpha = .05.$

The sample mean $\bar{x}$ is 13.87 is used in the test.

$$z = \frac{\bar{x} - \mu}{\sigma/\sqrt{n}} = \frac{13.87 - 13.2}{2.95/\sqrt{60}} = \frac{.67}{.381} = 1.76.$$

This is a two-tailed test with $\alpha = .05$. The critical region is $z > 1.96$ or $z < -1.96$. Since our $z = 1.76$ is not in the critical region, we fail to reject H_0. The evidence does not support rejection of $\mu = 13.2$.

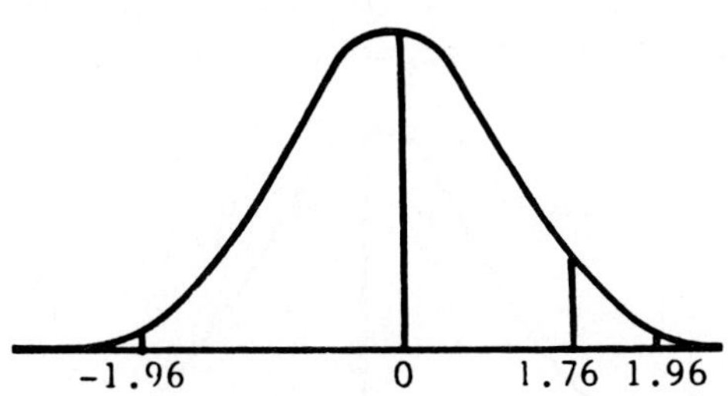

6-23 The claim is $\mu > 420$. The alternative is $\mu \le 420$. Our test will be

$H_0 : \mu \le 420$

$H_1 : \mu > 420 \qquad \alpha = .05.$

Computing z we have

$$z = \frac{385 - 420}{24/\sqrt{35}} = \frac{-35}{4.06} = -8.63.$$

This is a right-tailed test with $\alpha = .05$. The critical value is 1.645. Since -8.63 is not greater than the critical value of 1.645, we fail to reject H_0 and conclude that the evidence supports no improvement in reliability.

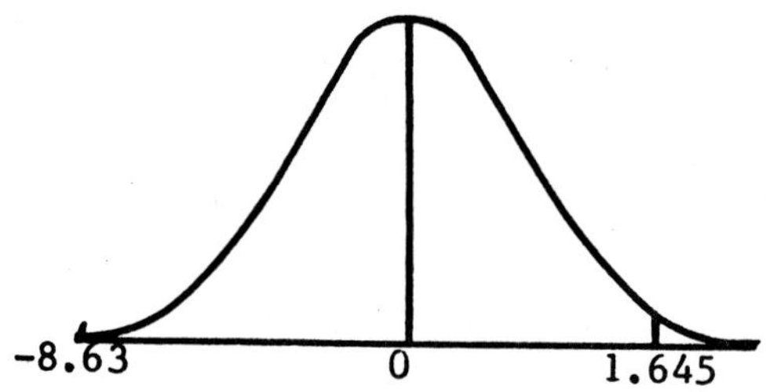

6-25 The claim that the mean is unchanged is a claim that $\mu = 22.6$. The alternative is $\mu \ne 22.6$. The test format is

$H_0 : \mu = 22.6$

$H_1 : \mu \ne 22.6 \qquad \alpha = .05$

Computing z we get

$$z = \frac{\bar{x} - \mu}{\sigma/\sqrt{n}} = \frac{20.1 - 22.6}{3.4/\sqrt{33}} = \frac{-2.5}{.592} = -4.22.$$

This is a two-tailed test with $\alpha = .05$. The critical region is more than 1.96 or less than -1.96. Our z value is less than -1.96 and hence we reject H_0. We conclude that the evidence supports the conclusion of $\mu \ne 22.6$.

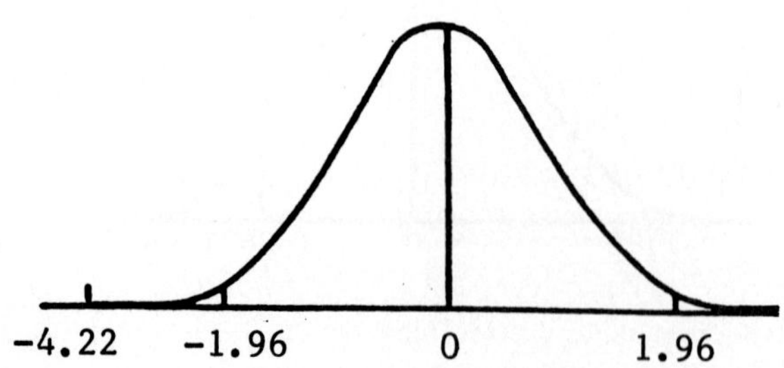

6-27 This exercise is solved in a manner similar to 6-25.

SOLUTIONS TO 6-2 B EXERCISES

6-29 For a mean above 100 we are claiming $\mu > 100$. Our alternative is $\mu \leq 100$. Then

$$H_0 : \mu \leq 100$$

$$H_1 : \mu > 100 \qquad \alpha = .03.$$

From the 34 data values we find $\bar{x} = 104.3$. The sample variance is

$$s^2 = \frac{n(\sum x^2) - (\sum x)^2}{n(n-1)} = \frac{34(375673) - (3545)^2}{34(33)} = 183.5.$$

Then the sample standard deviation is 13.5.

We form z by

$$z = \frac{104.3 - 100}{13.5/\sqrt{34}} = \frac{4.3}{2.32} = 1.85.$$

We have a right-tailed test with $\alpha = .03$. Our critical region will be $z > 1.885$. Since z is not in the critical region, we fail to reject H_0. There is no evidence to support $\mu > 100$.

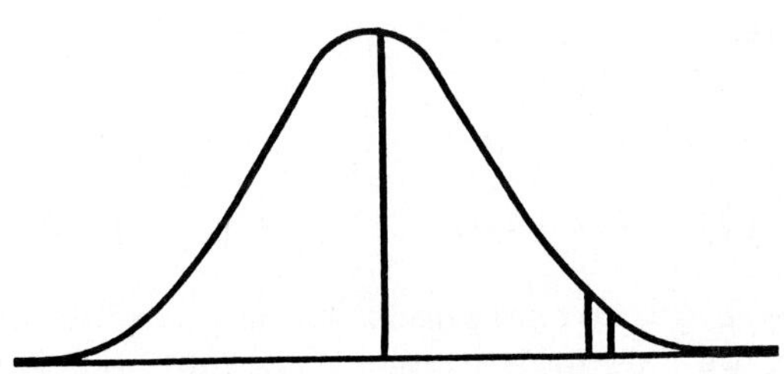

6-31 Our test is $\mu = 30$. Hence we have

$$H_0 : \mu = 30$$

$$H_1 : \mu \neq 30 \qquad \alpha = .04.$$

From the sample data $\bar{x} = 2474/87 = 28.4$, and

$$s^2 = \frac{n(\sum x^2 f) - (\sum xf)^2}{n(n-1)} = \frac{87(78098) - (2474)^2}{87(86)} = 90.06.$$

The sample standard deviation is s = 9.5.

The value of z is given by

$$z = \frac{\bar{x} - \mu}{s/\sqrt{n}} = \frac{28.4 - 30}{9.50/\sqrt{87}} = \frac{-1.6}{1.01} = -1.57.$$

This is a two-tailed test with $\alpha = .04$. The region where we do not reject the null hypothesis is $-2.055 \leq z \leq 2.055$. Since $z = -1.57$, we fail to reject H_0 and conclude that the evidence supports $\mu = 30$.

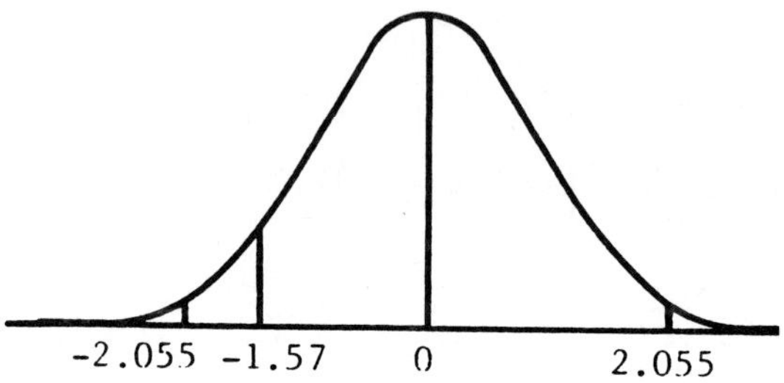

SOLUTIONS TO SECTION 6.3 - P VALUES

6-33 For a one-tailed test, since p = .03 is less than α .05, we reject the null hypothesis. For a two-tailed test, .03 is more than $\alpha/2$, and we would fail to reject H_0.

6-35 Since the p value is more than .10, we fail to reject H_0 for either one or two tails.

6-37 The p value for a z statistic of -1.40 is .0808. Since this is more than α = .05, we fail to reject H_0.

6-39 The p value for z = 1.94 is .0262. This is a one-tailed test with p value less than α = .05. We reject the null hypothesis.

6-41 A value of z = 1.66 corresponds to a p value of .0485. Since this is a two-tailed test, we double the p value and get .097. This is greater than α; hence we fail to reject H_0.

6-43 This is a two-tailed test. A value of z = -1.94 gives a p value of .0262. This is doubled to get .0524 which is greater than α = .05. As a result, we fail to reject H_0.

6-45 Our test is

$$H_0 : \mu \geq 100$$

$$H_1 : \mu < 100 \qquad \alpha = .05.$$

The z value is given by

$$z = \frac{\bar{x} - \mu}{\sigma/\sqrt{n}} = \frac{95 - 100}{15/\sqrt{45}} = \frac{-5}{2.24} = -2.23.$$

A z value of -2.23 corresponds to a p value of .0129. Since this is a one-tailed test and the p value is less than α, we reject H_0.

6-47 Our test is

$H_0 : \mu = 75.6$

$H_1 : \mu \neq 75.6 \qquad \alpha = .01.$

Computing z we get

$$z = \frac{\bar{x} - \mu}{\sigma/\sqrt{n}} = \frac{78.8 - 75.6}{12/\sqrt{81}} = \frac{3.2}{1.33} = 2.41.$$

A z value of 2.41 corresponds to a p value of .008. We double this to .016 since we have a two-tailed test. Since .016 is more than $\alpha = .01$, we fail to reject H_0.

6-49 A mean μ of at least 10.00 corresponds to $\mu \geq 10.00$. The alternative is $\mu < 10.00$. The form of the test is

$H_0 : \mu \geq 10$

$H_1 : \mu < 10 \qquad \alpha = .05.$

Computing z using the sample average we will have

$$z = \frac{\bar{x} - \mu}{\sigma/\sqrt{n}} = \frac{8.94 - 10}{8.53/\sqrt{60}} = \frac{-1.06}{1.10} = -.96.$$

A z value of -.96 corresponds to a p value of .1685. Since this is greater than α, we fail to reject H_0. We conclude that the evidence does not support $\mu < 10$.

6-51 Our claim is $\mu = 25$. Hence the test will be

$H_0 : \mu = 25$

$H_1 : \mu \neq 25 \qquad \alpha = .01.$

We determine z by

$$z = \frac{26.1 - 25}{13.08/\sqrt{950}} = \frac{1.1}{.424} = 2.59.$$

A z value of 2.59 corresponds to a p value of .0048. We double this because we have a two-tailed test. The result, .0096, is less than α and hence we reject H_0 and conclude that the evidence supports $\mu \neq 25$.

SOLUTIONS TO 6-3 B EXERCISES

6-53 We would still fail to reject H_0.

6-55 A p value of .2005 corresponds to a z value of .84. Then

$$z = .84 = \frac{\bar{x} - \mu}{\sigma/\sqrt{n}} = \frac{\bar{x} - 13.2}{2.95/\sqrt{60}} .$$

That is

$$.84 = \frac{\bar{x} - 13.2}{.38} .$$

Then $\bar{x} - 13.2 = (.84)(.38) = .32$,

or $\bar{x} = 13.2 + .32 = 13.52$.

SOLUTIONS TO SECTION 6.4 - t TEST

6-57 a. This is a two-tailed test with $\alpha = .05$ and 27 - 1 = 26 degrees of freedom. Consequently, the two critical t values are 2.0555 and -2.0555.

b. A right-tailed test with $\alpha = .10$ and 16 degrees of freedom has a critical t value of 1.3368.

c. A left-tailed test with $\alpha = .01$ and 5 degrees of freedom has a critical t value of -3.3649.

d. A two-tailed test with $\alpha = .01$ and 5 degrees of freedom has two critical t values. They are +4.0322 and -4.0322.

e. A left-tailed test with $\alpha = .01$ and 28 degrees of freedom has a critical t value of -2.4671.

6-59 Our test is

$H_0 : \mu \leq 10$

$H_1 : \mu > 10 \qquad \alpha = .05.$

We compute a t value because σ is unknown and $n < 30$. We get

$$t = \frac{\bar{x} - \mu}{s/\sqrt{n}} = \frac{11 - 10}{2/\sqrt{9}} = 1.5.$$

For a right-tailed test with 8 degrees of freedom and $\alpha = .05$, we have a critical t value of 1.8595. Since the computed value of t = 1.5 does not exceed the critical value, we fail to reject H_0, the null hypothesis.

6-61 Our test is

$H_0 : \mu \geq 98.6$

$H_1 : \mu < 98.6 \qquad \alpha = .025.$

Since σ is unknown and n < 30, the appropriate statistic is the t. We compute

$$t = \frac{\bar{x} - \mu}{s/\sqrt{n}} = \frac{98.2 - 98.6}{.8/\sqrt{18}} = \frac{-.4}{.189} = -2.12.$$

The critical t value for 17 degrees of freedom and α = .025 for this left-tailed test is -2.1098. Since the observed t is less than the critical value, we reject H_0. The evidence supports H_1.

6-63 Our test is

$H_0 : \mu = 75$

$H_1 : \mu \neq 75$ $\alpha = .05$.

Since σ is unknown and n < 30, we use the t test.

$$t = \frac{77.6 - 75}{5/\sqrt{15}} = \frac{2.6}{1.29} = 2.01.$$

The critical t values for 14 degrees of freedom, α = .05, and a two-tailed test are -2.1448 and +2.1448. Since our t value is between these two, we have no reason to reject H_0. The evidence supports $\mu = 75$.

6-65 The claim is $\mu < 154$. Then the alternative is $\mu \geq 154$. The test is

$H_0 : \mu \geq 154$

$H_1 : \mu < 154$ α .005.

We are testing for the mean with σ unknown and n < 30. We use the t statistic

$$t = \frac{\bar{x} - \mu}{s/\sqrt{n}} = \frac{141 - 154}{12/\sqrt{20}} = \frac{-13}{2.68} = -4.84.$$

It is a left-tailed test with α = .005 and 19 degrees of freedom. The critical value is -2.861. Since -4.84 is in the critical region to the left of -2.861, we reject H_0. Evidence does support $\mu < 154$.

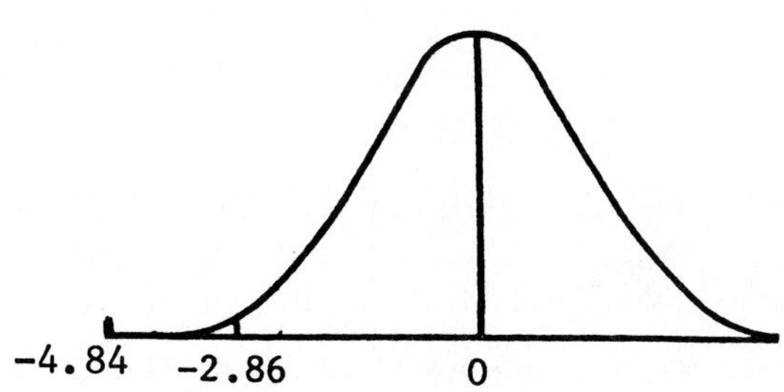

6-67 The claim is the mean is more than 500. The test will be

$H_0 : \mu \leq 500$

$H_1 : \mu > 500 \qquad \alpha = .05.$

We are testing for a mean with σ unknown and $n < 30$. Assuming the data is approximately normal, we use the t statistic. Then

$$t = \frac{524-500}{23/\sqrt{12}} = \frac{24}{6.64} = 3.61.$$

It is a right-tailed test with $\alpha = .05$ and 11 degrees of freedom. The critical value is 1.796. Since 3.61 is in the critical region greater than 1.796, we reject H_0. The evidence supports $\mu > 500$.

6-69 Our test is

$H_0 : \mu = 17850$

$H_1 : \mu \neq 17850 \qquad \alpha = .05.$

We are testing for a mean with σ unknown and $n < 30$. Assuming the data is approximately normal, we use the t statistic. Then

$$t = \frac{\bar{x} - \mu}{s/\sqrt{n}} = \frac{16735 - 17850}{978/\sqrt{5}} = \frac{-1115}{437} = -2.55.$$

It is a two-tailed test with $\alpha = .05$ and 4 degrees of freedom. The critical points are +2.776 and -2.776. Since 2.55 is not more than 2.5524, we accept H_0. The data does support a claim of $\mu \leq 17850$.

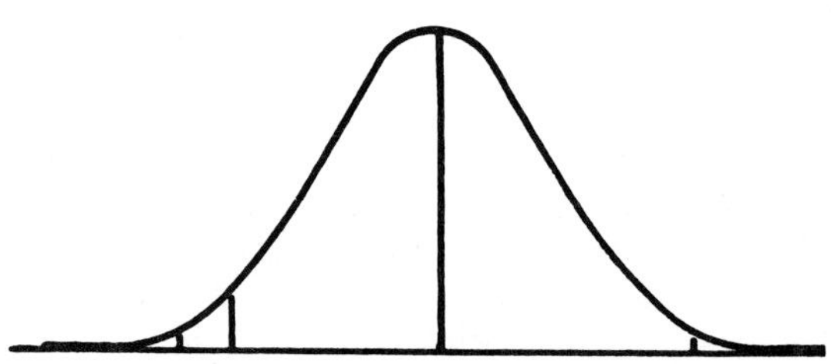

6-71 Our test is

$H_0 : \mu \leq 1500$

$H_1 : \mu > 1500 \qquad \alpha = .025.$

We are testing for a mean with σ unknown, and $n < 30$. Assuming that the data is approximately normal, we use the t statistic.

$$t = \frac{1756 - 1500}{843/\sqrt{25}} = \frac{256}{169} = 1.52.$$

It is a right tailed test with $\alpha = .025$ and 24 degrees of freedom. The critical value is 2.064. Since our t value is below this, it is not in the critical region. As a result, we fail to reject the claim that $\mu = 1500$.

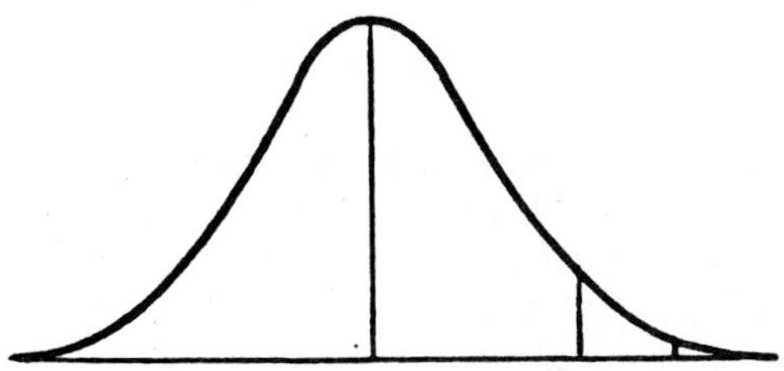

6-73 A two-tailed test similar in methodology to 6-69.

6-75 A one-tailed test similar in methodology to 6-71.

6-77 A mean more than 5 would signify $\mu > 5$. The alternative is $\mu \leq 5$. Our test is

$H_0 : \mu \leq 5$

$H_1 : \mu > 5 \qquad \alpha = .10.$

We are testing for a mean with σ unknown and $n < 30$. Assuming the data is approximately normal, we use a t test. Then

$$t = \frac{\bar{x} - \mu}{s/\sqrt{n}} = \frac{5.15 - 5}{1.68/\sqrt{28}} = \frac{.15}{.317} = .47.$$

This is a right-tailed test with $\alpha = .10$ and 27 degrees of freedom. The critical region is to the right of 1.28. Since .47 is not in that region, we fail to reject H_0 and conclude that the data supports $\mu \leq 5$.

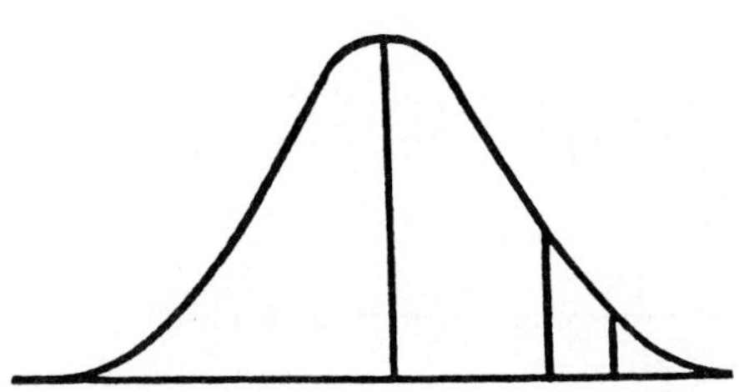

6-79 Our claim is $\mu \neq 41.9$. The test will be

$H_0 : \mu = 41.9$

$H_1 : \mu \neq 41.9 \qquad \alpha = .01.$

We are testing for a mean with σ unknown and $n < 30$. Assuming the data is approximately normally distributed, we use the t test. Then

$$t = \frac{31 - 41.9}{10.5/\sqrt{15}} = \frac{-10.9}{2.71} = -4.02.$$

This is a two-tailed test with $\alpha = .01$ and 14 degrees of freedom. The critical values are ± 2.977. Since -4.02 is less than -2.977, we reject H_0. The evidence supports the claim of lower mean.

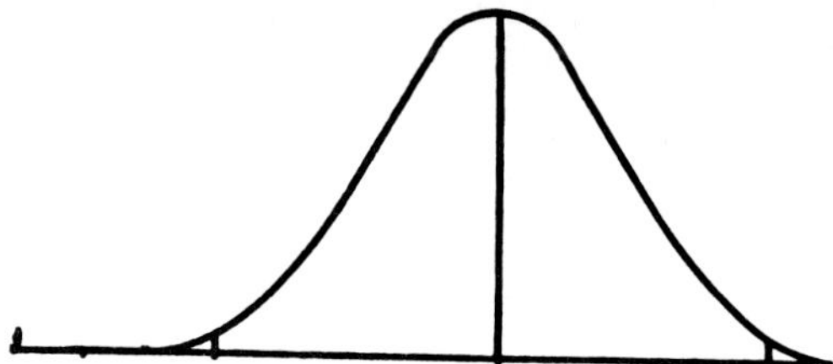

6-81 Solved like 6-69.

SOLUTIONS TO 6-4 B EXERCISES

6-83 The claim of above average class is a claim that $\mu > 75$. The form of the test is

$H_0 : \mu \leq 75$

$H_1 : \mu > 75 \qquad \alpha = .05.$

This is a test about a mean with σ unknown and $n < 30$. Assuming the data is approximately normal, we use the t test. From the data we compute

$$\bar{x} = \frac{1561}{20} = 78.05.$$

Also

$$s^2 = \frac{n\sum x^2 - (\sum x)^2}{n(n-1)} = \frac{20(122431) - (1561)^2}{20(19)} = 31.3, \text{ so } s = 5.6.$$

$$t = \frac{\bar{x} - \mu}{s/\sqrt{n}} = \frac{78.05 - 75}{5.60/\sqrt{20}} = \frac{3.05}{1.25} = 2.44.$$

This is a right-tailed test with 19 degrees of freedom, and $\alpha = .05$. The critical value is 1.7291. Since $t = 2.44$ does exceed this, we reject H_0. The data supports this as an exceptional class.

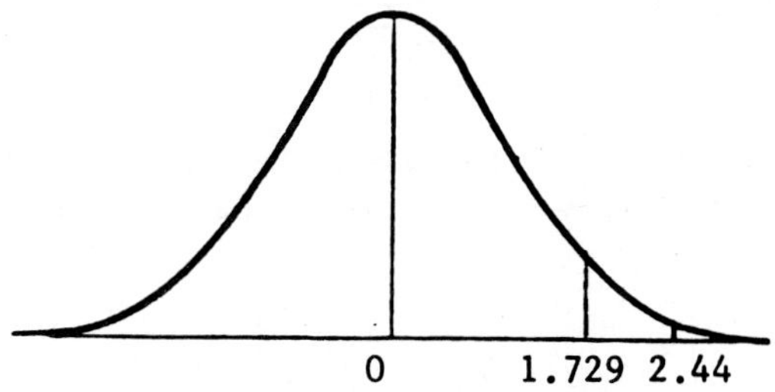

6-87 The critical z score is 1.645 for a right tailed test with α = .05. Using DF = 9 degrees of freedom, A = 1.645(1.027) = 1.68. Then A^2/DF = .3136. Then e to that power is 1.368. Then t is the square root of 9(1.368 - 1) = 3.312. That is t = 1.82. From the table the correct value of t is 1.833.

SOLUTIONS TO SECTION 6.5 - TESTS OF PROPORTIONS

6-89 Our test is

$$H_0 : p = .3$$

$$H_1 : p \neq .3 \qquad \alpha = .05.$$

Because we are testing proportions, the test statistic is z. By the formula

$$z = \frac{x - np}{\sqrt{npq}} = \frac{45 - (100)(.3)}{\sqrt{100(.3)(.7)}} = \frac{15}{4.58} = 3.28.$$

It is a two-tailed test with α = .05. The critical z values are +1.96 and -1.96. Since 3.28 is more than 1.96, we reject H_0 that p = .3. The data supports $p \neq .3$.

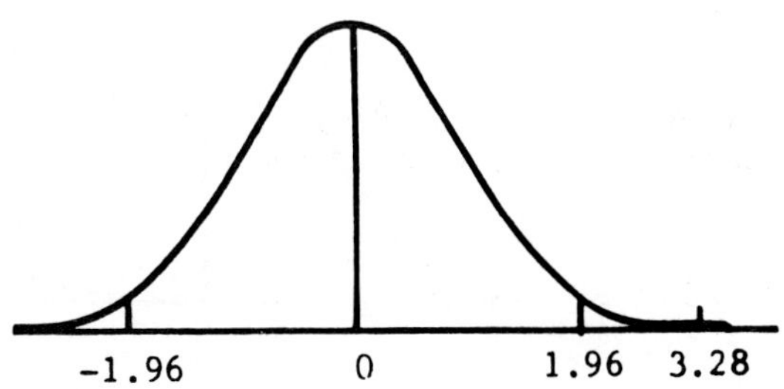

6-91 The claim is that p < .7. The alternative will be p ≥ .7. Then the test is

$$H_0 : p \geq .7$$

$$H_1 : p < .7 \qquad \alpha = .01.$$

Because we are testing proportions, the appropriate test statistic is z. Then

$$z = \frac{x - np}{\sqrt{npq}} = \frac{31 - (50)(.7)}{\sqrt{50(.7)(.3)}} = \frac{-4}{\sqrt{10.5}} = \frac{-4}{3.24} = -1.23.$$

For a one-tailed test with $\alpha = .01$, the z value is -2.325. Since -1.23 is not less than the critical value, we fail to reject the hypothesis that $p \geq .7$.

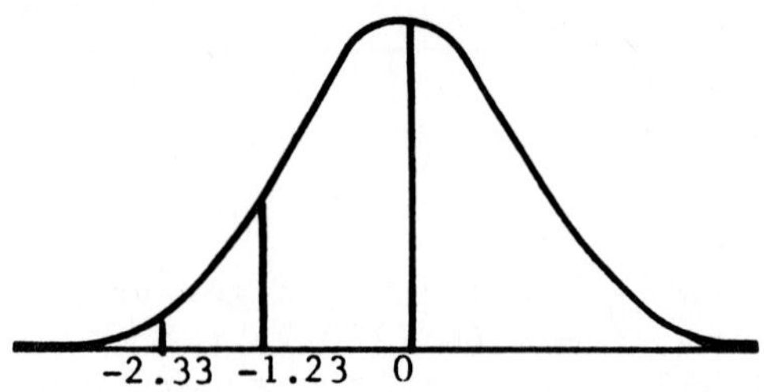

6-93 For a more than type test, we are testing $p > .25$. The test is then

$H_0 : p \leq .25$

$H_1 : p > .25 \qquad \alpha = .02.$

We use the z statistic when testing proportions. With $x = (.33)(1400) = 462$,

$$z = \frac{x - np}{\sqrt{npq}} = \frac{462 - (1400)(.25)}{\sqrt{(1400)(.25)(.75)}} = \frac{112.0}{16.20} = 6.91.$$

For a right-tailed test with $\alpha = .02$, the critical z value is 2.05. Since 6.91 is more than this, it is in the critical region and hence we reject H_0. The evidence supports a proportion more than .25.

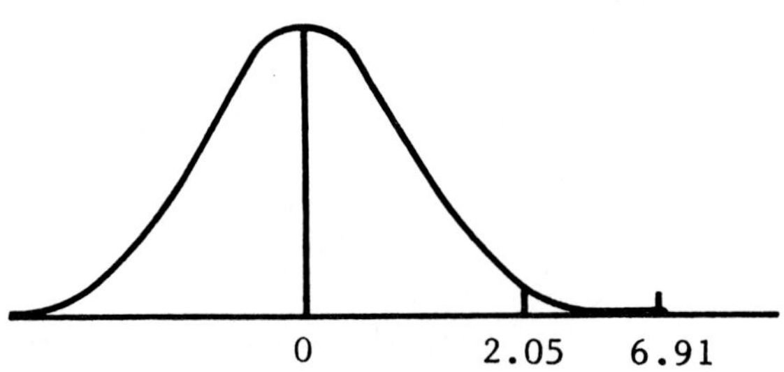

6-95 We test

$H_0 : p = .71$

$H_1 : p \neq .71 \qquad \alpha = .10.$

Using the z statistic we have

$$z = \frac{x - np}{\sqrt{npq}} = \frac{370 - (500)(.71)}{\sqrt{(500)(.71)(.29)}} = \frac{15}{10.1} = 1.48.$$

For a two-tailed test with $\alpha = .10$, the critical values are ±1.645. Since 1.48 does not exceed 1.645, it is not in the critical region, and we fail to reject H_0. The data supports the conclusion that the proportion is .71.

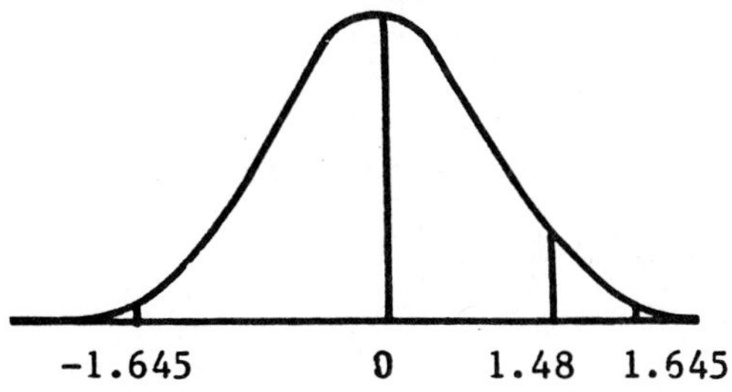

6-97 This exercise is solved in the same manner as 6-95.

6-99 This exercise is solved in the same manner as 6-95.

6-101 This exercise is solved in the same manner as 6-95.

6-103 Our test is a right-tailed test since the claim is $p > 1/3$.

$H_0 : p \leq 1/3$

$H_1 : p > 1/3.$

The z value is

$$z = \frac{x - np}{\sqrt{npq}} = \frac{120-(324)(1/3)}{\sqrt{(324)(.33)(.67)}} = \frac{12}{8.46} = 1.42.$$

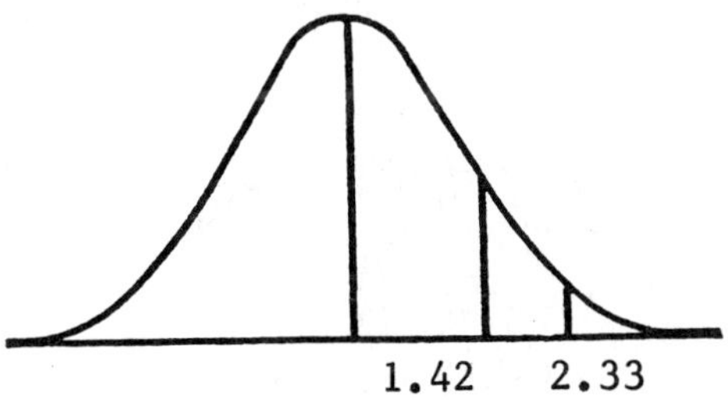

The critical value is 2.33. We do not reject H_0.

6-105 We are concerned with $p > .03$. The test will then be

$H_0 : p \leq .03$

$H_1 : p > .03 \qquad \alpha = .05.$

Using the z statistic, we have

$$z = \frac{x - np}{\sqrt{npq}} = \frac{22 - (500)(.03)}{\sqrt{500(.03)(.97)}} = \frac{7}{3.81} = 1.84.$$

For a one-tailed test with $\alpha = .05$ the critical z value is 1.64. Since the observed z of 1.84 exceeds this, we reject H_0 and conclude that evidence supports the system being out of control.

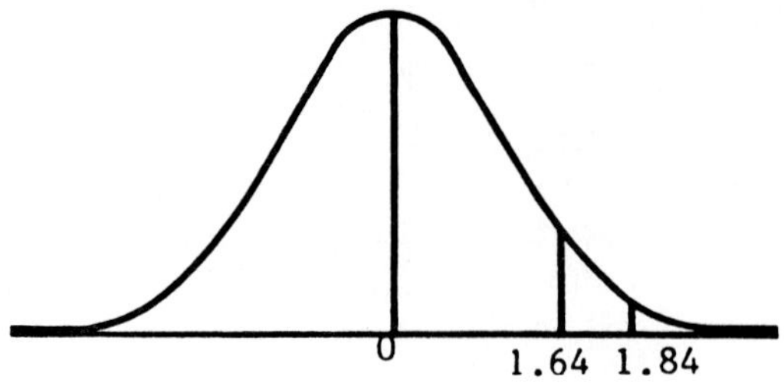

6-107 Our claim is $p < 1/2$. Then the form of the test is

$$H_0 : p \geq 1/2$$

$$H_1 : p < 1/2 \qquad \alpha = .01.$$

For testing proportions, the appropriate statistic is a z test. Then

$$z = \frac{x - np}{\sqrt{npq}} = \frac{156 - (1/2)(400)}{\sqrt{(400)(1/2)(1/2)}} = \frac{-44}{10} = -4.4.$$

In this case we had $x = (.39)(400) = 156$. For a left-tailed z test the critical value is -2.33. Since -4.6 is less than the critical value, we reject H_0 and conclude that evidence supports the claim of fewer than 1/2.

SOLUTIONS TO 6-5 B EXERCISES

6-109 Because this test is p no more than some given p_0, our test would be

$$H_0 : p \leq p_0$$

$$H_1 : p > p_0 \qquad \alpha = .05.$$

A right-tailed z test with $\alpha = .05$ has a critical z value of 1.64. From our equation we need $\hat{p}$ where

$n = 500, \qquad np = 15$. Then,

$$1.64 = \frac{15 - 500p}{\sqrt{500p(1-p)}}$$

or

$$1.64\sqrt{500(p - p^2)} = 15 - 500p.$$

Squaring both sides yields

$$2.69(500(p - p^2)) = 225 - 15000p + 250000p^2$$

or

$$1345p - 1345p^2 = 225 - 15000p + 250000p^2.$$

Then

$$251345p^2 - 16345p + 225 = 0.$$

Using the quadratic equation

$$p = \frac{16345 \pm \sqrt{(16455)^2 - 4(225)(251345)}}{2(251345)}$$

Using only the - sign to get the smaller, p = .0197.

6-111 Solved like 6-109.

SOLUTIONS TO SECTION 6.6 - TESTS OF VARIANCES

6-113 a. For a two-tailed test with α = .05 and 19 degrees of freedom, χ^2 = 8.907 and χ^2 = 32.852 are the two critical values.

b. For a left-tailed test with α = .05 and 19 degrees of freedom, the critical χ^2 value is 10.117.

c. For a two-tailed test, α = .05 and 22 degrees of freedom, the critical values are χ^2 = 8.643 and χ^2 = 42.796.

d. For a right-tailed test, 22 degrees of freedom, and α = .05, the critical χ^2 value is 40.289.

e. For 14 degrees of freedom, a left-tailed test and α = .005, the critical χ^2 is 4.075.

6-115 Since we claim that $\sigma^2 > 100$, the form of the test would be

$H_0 : \sigma^2 \leq 100$

$H_1 : \sigma^2 > 100$

The value of χ^2 is

$$\chi^2 = \frac{(n-1)s^2}{\sigma^2} = \frac{(26)(194)}{100} = \frac{5044}{100} = 50.4.$$

There are n - 1 = 27 - 1 = 26 degrees of freedom. Because the form of H_1 is σ^2 greater than 100, we will only want to reject H_0 if χ^2 is too large. With α = .05, the critical χ^2 value is 38.885. Since the observed χ^2 = 50.4 exceeds that, we reject H_0 in favor of H_1 concluding that the data supports a conclusion of $\sigma > 100$.

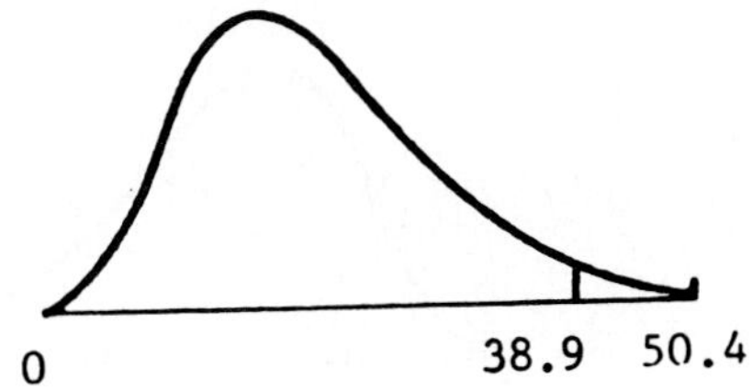

6-117 Our test is

$$H_0 : \sigma^2 \geq 416$$

$$H_1 : \sigma^2 < 416 \qquad \alpha = .05.$$

Our test statistic is χ^2 and so

$$\chi^2 = \frac{(n-1)s^2}{\sigma^2} = \frac{(16)(247)}{416} = 9.5.$$

A left-tail test with 16 degrees of freedom and α = .05 has a critical value of 7.962. Since the observed χ^2 is not to the left of this, we fail to reject H_0 that $\sigma^2 \geq 416$.

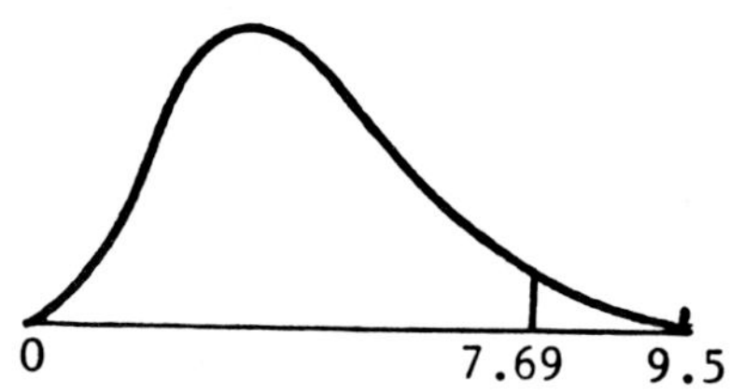

6-119 Our test is

$$H_0 : \sigma^2 = 2.38$$

$$H_1 : \sigma^2 \neq 2.38 \qquad \alpha - .01.$$

The appropriate statistic is χ^2. We compute

$$\chi^2 = \frac{(n-1)s^2}{\sigma^2} = \frac{(11)(5)}{2.38} = 23.11.$$

For a two-tailed test with α = .01 and 11 degrees of freedom, the two critical values are χ^2 = 2.6 and χ^2 = 26.757. Since our observed χ^2 is between these, it is not in the critical region, so we fail to reject H_0 that σ^2 = 2.38.

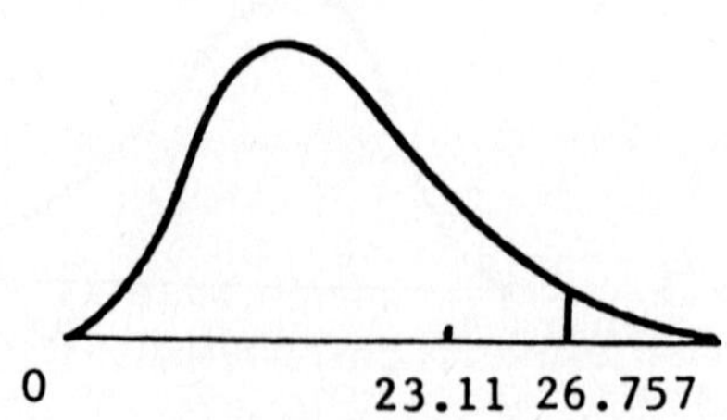

6-121 Our test is

$H_0 : \sigma = 10$

$H_1 : \sigma \neq 10 \qquad \alpha = .01.$

The appropriate test is the χ^2 test on variances. Then

$$\chi^2 = \frac{(n-1)s^2}{\sigma^2} = \frac{(17)(210.25)}{100} = 35.74.$$

For a two-tailed χ^2 test with 17 degrees of freedom and $\alpha = .01$, the critical values are 5.697 and 35.718. Since our χ^2 value exceeds 35.718, we reject H_0. The data supports H_1 that $\sigma \neq 10$.

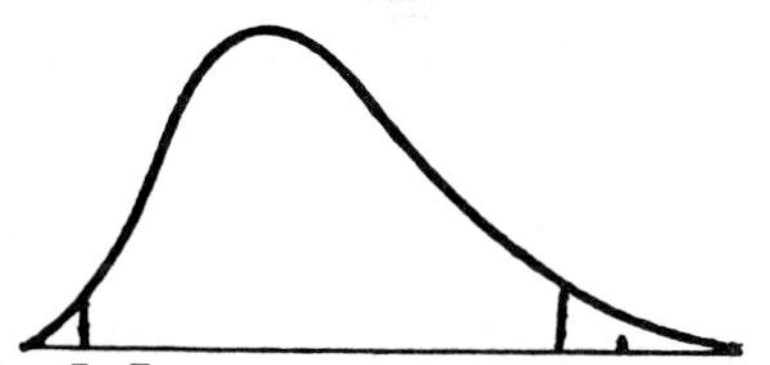

6-123 This is a right-tailed test similar to 6-115. The critical value is χ^2 = 101.879 for 80 degrees of freedom. Our χ^2 of (80)(12.25)/9 = 108.9 exceeds the critical value. Hence we would reject H_0.

6-125 Our claim is $\sigma > 2.0$. Hence the test is

$H_0 : \sigma \leq 2.0$

$H_1 : \sigma > 2.0 \qquad \alpha = .025.$

We use a χ^2 test on the variances. Hence

$$\chi^2 = \frac{(n-1)s^2}{\sigma^2} = \frac{11(2.85)^2}{2^2} = 22.34.$$

For a right-tailed test with 11 degrees of freedom we have a critical χ^2 value of 21.92. Since our observed χ^2 value exceeds this amount, we reject H_0 and conclude that the data supports $\sigma > 2$.

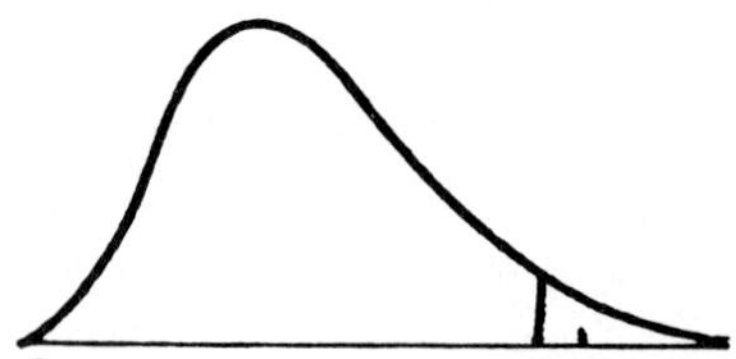

6-127 The claim is that σ^2 is less than $(.15)^2$. The test is

$H_0 : \sigma^2 \geq .0225$

$H_1 : \sigma^2 < .0225 \qquad \alpha = .05.$

For variances,

$$\chi^2 = \frac{(n-1)s^2}{\sigma^2} = \frac{70(.0144)}{.0225} = 44.8.$$

For a left-tailed test and $\alpha = .05$, the critical χ^2 value is 51.739. Since that is less than the critical value, we reject H_0. The data supports $\sigma^2 < .0225$.

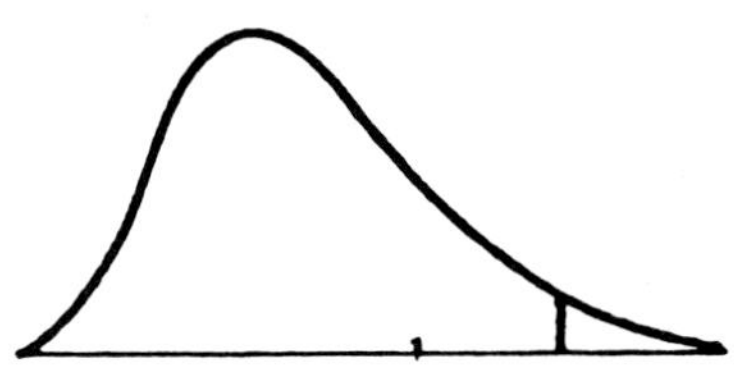

6-129 This is a one-tailed test similar to 6-123.

6-131 Our test is

$H_0 : \sigma \leq 50$

$H_1 : \sigma > 50 \qquad \alpha = .025.$

For tests on variances we use a χ^2 test.

$$\chi^2 = \frac{(n-1)s^2}{\sigma^2} = \frac{21(65^2)}{(50)^2} = 35.49.$$

For a one-tailed χ^2 test with 21 degrees of freedom and $\alpha = .025$, the critical value is $\chi^2 = 35.479$. Since our observed χ^2 value is more than 35.479, we reject H_0. The evidence supports $\sigma > 50$.

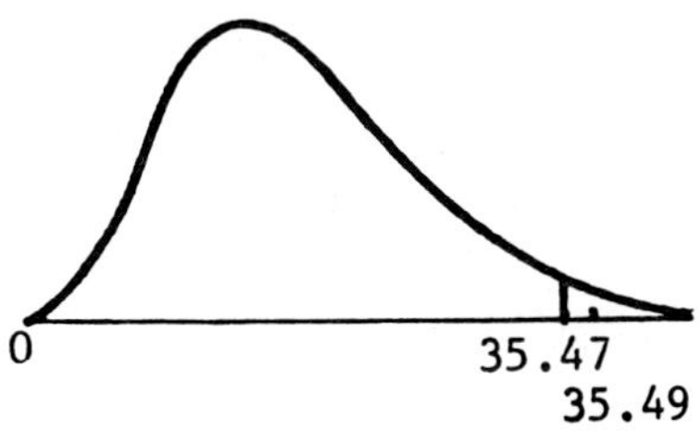

6-133 This exercise is solved like previous exercises.

6-135 We are testing

$H_0 : \sigma^2 \geq 4$

$H_1 : \sigma^2 < 4 \qquad \alpha = .025.$

From the data we compute S^2 as

$$s^2 = \frac{n(\sum x^2) - (\sum x)^2}{n(n-1)} = \frac{(12)(13121.66) - (396.6)^2}{12(11)} = 1.275 \text{ and } s = 1.13.$$

Our χ^2 value is $\chi^2 = \frac{11(1.275)}{4} = 3.51.$

The one-tailed critical value for 11 degrees of freedom is 3.816. Since our observed value is less than this, we reject H_0. The data supports a conclusion of $\sigma^2 < 4$.

SOLUTIONS TO 6-6 B EXERCISES

6-137 a. For a two-tailed test, $z = \pm 1.96$. We have k = 100 degrees of freedom. The two χ^2 values are

$$\chi^2 = \frac{1}{2}(1.96 + \sqrt{2(100) - 1})^2 = 129.07$$

and

$$\chi^2 = \frac{1}{2}(-1.96 + \sqrt{2(100) - 1})^2 = 73.77.$$

From table A5 the correct values are 74.22 and 129.561. The two sets are quite close.

b. As in a.), the two values for z are ± 1.96. Here we have k = 149 degrees of freedom. The two χ^2 values are

$$\chi^2 = \frac{1}{2}(1.96 + \sqrt{2(149) - 1})^2 = 184.2$$

and

$$\chi^2 = \frac{1}{2}\,(-1.96 + \sqrt{2(149) - 1})^2 = 116.6.$$

6-139 Computing s using the formula for grouped data of chapter 2, we proceed as in the other exercises of this section.

SOLUTIONS TO REVIEW EXERCISES

6-141 a. Tests on proportions are z tests. The critical value for this left-tailed test is -1.64.

b. This is a test of means. Since n exceeds 30, the normal table would be used even if σ is not known. Therefore, the critical value for this right tailed test is 2.33.

c. The answer here would depend upon whether or not σ is known. If σ is known the answers are ± 2.57. If not, then from the t table the critical values are ± 3.106.

d. From the χ^2 table the critical value is 10.856.

e. From the χ^2 table the critical values are 16.047 and 45.722.

6-143 a. $H_0 : \mu \geq 20.0$.

b. It is a left-tailed test since the alternate hypothesis is $H_1 : \mu < 20.0$.

c. The type I error in this case is a conclusion that H_0 is to be rejected when in fact H_0 is true and μ is in fact ≥ 20.

d. The type II error occurs when we fail to reject H_0 but H_0 is false. In this case that would mean $\mu < 20$.

e. The magnitude of the α level, $\alpha = .01$.

6-145 Our test is

$H_0 : p = .20$

$H_1 : p \neq .20 \qquad \alpha = .05.$

For a test of proportions we use

$$z = \frac{x - np}{\sqrt{npq}} = \frac{132 - 600(.20)}{\sqrt{600(.20)(.80)}} = \frac{12}{9.8} = 1.22.$$

The critical values for a two-tailed z test with $\alpha = .05$ are $\pm$ 1.96. Since our observed z is between these two, we fail to reject H_0. We conclude that the data supports $p = .20$.

6-147 The claim is

$H_0 : \sigma = 4$

$H_1 : \sigma \neq 4 \qquad \alpha = .05.$

We use the χ^2 test for variances here. We compute

$$\chi^2 = \frac{(n-1)s^2}{\sigma^2} = \frac{(19)(2.8)^2}{(4)^2} = 9.31.$$

This is a two-tailed test with $\alpha = .05$ and 19 degrees of freedom. The critical χ^2 valuies are 8.907 and 32.852. Since the observed χ^2 is between these two, we fail to reject H_0. The evidence supports $\sigma = 4$.

6-149 Our claim is $\mu = 10.0$. The form of the test will be

$H_0 : \mu = 10$

$H_1 : \mu \neq 10 \qquad \alpha = .05.$

Since σ is known and we are testing about means, the appropriate statistic is

$$z = \frac{\bar{x} - \mu}{\sigma/\sqrt{n}} = \frac{8.2 - 10}{6/\sqrt{100}} = \frac{-1.8}{.6} = -3.$$

For a two-tailed z test with $\alpha = .05$, the critical values are $\pm$ 1.96. Since -3 is not between these, we reject H_0. The data suports a conclusion of $\mu \neq 10$.

6-151 Our test is

$H_0 : p \leq 1/5$

$H_1 : p > 1/5 \qquad \alpha = .05.$

For a test of proportions we have .27 x 200 = 54 and compute

$$z = \frac{x - np}{\sqrt{npq}} = \frac{54 - 40}{\sqrt{(200)(1/5)(4/5)}} = \frac{14}{5.66} = 2.47.$$

The critical z value for this right-tailed test with $\alpha = .05$ is 1.64. Since our observed z exceeds this, we reject H_0. The data supports $p > 1/5$.

6-153 Our claim is $\mu < 5$. Hence the form of the test is

$H_0 : \mu \geq 5$

$H_1 : \mu < 5 \qquad \alpha = .01.$

Since σ is unknown and n = 36, we use the z test with s instead of σ. Then

$$z = \frac{\bar{x} - \mu}{s/\sqrt{n}} = \frac{4.13 - 5}{1.91/\sqrt{35}} = \frac{-.87}{.323} = -2.69.$$

For a left-tailed test on means, the critical value of z is -2.33. Since our observed value of z is less than the critical value, we reject H_0. The data supports a conclusion of $\mu < 5$.

6-155 Our test will be

$H_0 : \sigma^2 \leq 6410$

$H_1 : \sigma^2 > 6410$.

Using our χ^2 test we compute

$$\chi^2 = \frac{(n-1)s^2}{\sigma^2} = \frac{(59)(8464)}{(6410)} = 77.9.$$

To determine critical value we can use the formula developed in problem 6-137. Another method would be to observe that the critical value is less than the right-tailed $\alpha = .1$ critical value for 60 degrees of freedom. Therefore, the critical value is less than 74.397. Since 77.9 exceeds any such number, it is in the critical region. Consequently we reject H_0. The evidence supports a conclusion of $\sigma^2 > 6410$.

6-157 Our test is

$H_0 : p \leq .50$

$H_1 : p > .50 \qquad \alpha = .05$.

We compute

$$z = \frac{x - np}{\sqrt{npq}} = \frac{55 - 40}{4.47} = 3.35.$$

For a right-tailed z test with $\alpha = .05$, the critical z value is 1.645. Since the observed z does exceed the critical value, we fail to accept H_0. The data supports the claim of more than half.

6-159 Our test is

$H_0 : \mu = 100$

$H_1 : \mu \neq 100 \qquad \alpha = .01$

We compute from the sample data

$$\bar{x} = \frac{\sum x}{n} = \frac{1470}{15} = 98.$$

Also

$$s^2 = \frac{n\sum x^2 - (\sum x)^2}{n(n-1)} = \frac{15(145050) - 1470)^2}{15(14)} = 70.7.$$

Then $s = 8.41$. Since σ is unknown and $n < 30$, we use a t test. Then

$$t = \frac{\bar{x} - \mu}{s/\sqrt{n}} = \frac{98 - 100}{8.41/\sqrt{15}} = \frac{-2}{2.17} = -.92.$$

For a two-tailed t test with 14 degrees of freedom and $\alpha = .01$, the critical values are ± 2.977. Since the observed t value of -.92 is between these, we fail to reject H_0. The data supports a conclusion of $\mu = 100$.

CHAPTER 7. ESTIMATES AND SAMPLE SIZES

SOLUTIONS TO SECTION 7.2 ESTIMATES AND SAMPLE SIZES OF MEANS

7-1 a. For $\alpha = .05$, then $\alpha/2 = .025$ and using the z table, $z(\alpha/2) = 1.96$.

c. Since the confidence level is 96%, then $\alpha = .04$ and $\alpha/2 = .02$. Using the z table, $z(\alpha/2) = 2.05$.

e. With 15 scores, there are 15 - 1 = 14 degrees of freedom. This t value would be the two tailed .01 column. Consequently, $t(\alpha/2) = 2.977$.

7-3 In this case σ is known. As a result, the formula is $E = (z(\alpha/2))(\sigma/\sqrt{n})$. Since $\alpha = .05$, $z(\alpha/2) = 1.96$. The maximum error of the estimate E is $E = 1.96(15/\sqrt{100}) = ((1.96)(15))/10$ or 2.94.

7-5 As in 7-3, σ is known, and thus the same formula for maximum error of the estimate is used with $z(\alpha/2) = 2.575$. Then

$$E = z(\alpha/2)(\sigma/\sqrt{n}) = 2.575(40/\sqrt{25}) = 20.6.$$

7-7 Although σ is unknown, the sample size is greater than 30. As a result, we can use the z table instead of the t table. The appropriate formula would be

$$E = z(\alpha/2)(s/\sqrt{n}) = 2.575(15/\sqrt{100}) = 3.863.$$

7-9 To find the confidence interval we evaluate

$$\bar{X} - E < \mu < \bar{X} + E$$

where E is the maximum error of the estimate. The appropriate formula for E is based upon the fact that σ is known. Then

$$E = z(\alpha/2)(\sigma/\sqrt{n}) = 1.96(5/\sqrt{36}) = 9.8/6 = 1.63.$$

The confidence interval is

$$70.4 - 1.63 < \mu \quad 70.4 + 1.63$$

or

$$68.77 < \mu < 72.03.$$

7-11 The confidence interval is given by

$$\bar{x} - E < \mu < \bar{x} + E$$

for the appropriate value of E. As in 7-9, σ is known and therefore

$$E = z(\alpha/2)(\sigma/\sqrt{n}) = (2.575)(2/\sqrt{100}) = .515.$$

The 99% confidence interval would then be

$$98.6 - .515 < \mu < 98.6 + .515$$

or

$$98.085 < \mu < 99.115.$$

7-13 To evaluate the confidence interval we use

$$\bar{x} - E < \mu < \bar{x} + E.$$

Since σ is not known, but the sample size is more that 30, then the error in the estimate E is given by

$$E = z(\alpha/2)(s/\sqrt{n}) = (1.96)(.75/\sqrt{231}) = .10.$$

Consequently the confidence interval is

$$\bar{x} - E < \mu < \bar{x} + E$$

or

$$3.94 - .10 < \mu < 3.94 + .10$$

or

$$3.84 < \mu < 4.04.$$

7-15 As above, the formula for the confidence interval is

$$\bar{x} - E < \mu < \bar{x} + E.$$

In this problem the standard deviation is estimated using the sample s. The sample size exceeds 30, and, as a result, the appropriate formula uses a z score and

$$E = z(\alpha/2)(s/\sqrt{n}) = 2.57(1.68/\sqrt{4400}) = .065.$$

Then the 99% confidence interval is

$$5.15 - .065 < \mu < 5.15 + .065$$

or

$$5.08 < \mu < 5.22.$$

7-17 The maximum error in the estimate is based upon student's t because σ is unknown and the sample size is under 30. There are n - 1 = 29 - 1 = 28 degrees of freedom. Then $t(\alpha/2) = 2.048$, and

$$E = t(\alpha/2)(s/\sqrt{n}) = 2.048(17.66/\sqrt{29}) = 6.72.$$

The 95% confidence interval is

$$\bar{x} - E < \mu < \bar{x} + E$$

or

$$26.84 - 6.72 < \mu < 26.84 + 6.72$$

or

$$20.12 < \mu < 33.56.$$

7-19 From the information given in the problem, it is apparent that the standard deviation was based upon information from the sample. That is, σ is unknown and s = .004. Since the sample size n = 25 is less than 30, student's t is the appropriate statistic. $t(\alpha/2) = 2.064$ based upon $\alpha = .05$ and 24 degrees of freedom. The maximum error of the estimate is

$$E = t(\alpha/2)(s/\sqrt{n}) = 2.492(.004/\sqrt{25}) = .002.$$

The 95% confidence interval is

$$\bar{x} - E < \mu < \bar{x} + e$$

or

$$.996 - .002 < \mu < .996 + .002$$

or

$$.994 < \mu < .998.$$

7-21 Using the sample size formula 7-1 since σ is known,

$$n = \left[\frac{z(\alpha/2)\sigma}{E}\right]^2 = \left[\frac{(1.96)(16)}{3}\right]^2 = (10.45)^2 = 109.2.$$

We round this up to 110 samples.

7-23 Using the sample size formula with σ known,

$$n = \left[\frac{z(\alpha/2)\sigma}{E}\right]^2 = \left[\frac{(2.575)(15)}{.5}\right]^2 = (77.25)^2 = 5967.6.$$

We round this up to 5968 samples.

7-25 Although σ is not known, the sample size is larger than 30 so the maximum error of the estimate is

$$E = z(\alpha/2)(s/\sqrt{n}) = (2.33)(7.3/\sqrt{320}) = .95.$$

Here $\alpha = .02$ and $z(\alpha/2) = 2.33$. Then the 98% confidence interval is

$$\bar{x} - E < \mu < \bar{x} + E$$

or

$$28.8 - .95 < \mu < 28.8 + .95$$

or

$$27.85 < \mu < 28.75.$$

7-27 Since σ is known to be 2.5, the maximum error of the estimate for 95% confidence is

$$E = z(\alpha/2)(\sigma/\sqrt{n}) = (1.96)(2.5/\sqrt{7}) = 1.85.$$

Then the 95% confidence interval is

$$14.7 - 1.85 < \mu < 14.7 + 1.85$$

or

$$12.85 < \mu < 16.55.$$

7-29 When σ is unknown and the sample size is less than 30, we use student's t. The maximum error of the estimate is

$$E = t(\alpha/2)(s/\sqrt{n}) = (2.064)(5.5/\sqrt{25}) = 2.27$$

where the t value is based upon 24 degrees of freedom. The 95% confidence interval is then

$$\bar{x} - E < \mu < \bar{x} + E$$

$$68.5 - 2.27 < \mu < 68.5 + 2.27$$

$$66.23 < \mu < 70.77.$$

7-31 Because the standard deviation is estimated from the sample, we have σ unknown. Since in addition n is less than 30, we use student's t for the maximum error of the estimate. By formula, using $\alpha = .01$ and $n - 1 = 18 - 1 = 17$ degrees of freedom

$$E = t(\alpha/2)(s/\sqrt{n}) = 2.898(3.4/\sqrt{18}) = (9.8532/4.243) = 2.32.$$

The 99% confidence interval will then be

$$16.2 - 2.32 < \mu < 16.2 + 2.32$$

or

$$13.88 < \mu < 18.52.$$

7-33 Using the sample size formula

$$n = \left[\frac{z(\alpha/2)\sigma}{E}\right]^2 = \left[\frac{(2.57)(2.80)}{.5}\right]^2 = (14.39)^2 = 208$$

when rounded up to the next whole sample.

7-35 In this example we wish to be 98% confident, and, as a result, $\alpha = .02$. Then $z(\alpha/2) = 2.33$. The sample size needed is given by

$$n = \left[\frac{z(\alpha/2)\sigma}{E}\right]^2 = \left[\frac{(2.33)(10)}{3}\right]^2 = (7.77)^2 = 61.$$

rounding up to the nearest whole sample.

7-37 Using the data we compute the mean as

$$\bar{x} = \frac{\Sigma x}{n} = \frac{35.58}{10} = 3.56. \quad \text{Also } s = 1.23.$$

Using students t with 9 degrees of freedom, we have

$E = t(\alpha/2)(s/\sqrt{n}) = 2.262(1.23/3.16) = .88.$

Then the 95% confidence interval is given by

$x - E < \mu < x + E$

or

$3.56 - .88 < \mu < 3.56 + .88$

or

$2.68 < \mu < 4.44.$

7-39 In order to specify the interval estimate, we shall choose to find a 95% confidence interval. Although σ is not known, the sample size is 36. As a result, we shall use $z(\alpha/2)$ in our computation of the maximum error of the estimate. We use the data given to compute the mean and standard deviation. We have $\bar{x} = 52492$ and $s = 29112$. These were computed using the formulas of chapter 2. We determine the standard error of the estimate as for a 95% confidence interval as

$$E = (Z(\alpha/2))(s/\sqrt{n}) = 1.96\ (29112/\sqrt{36}) = 9510.$$

Then the 95% confidence interval is given by

$$52492 - 9510 < \mu < 52492 + 9510$$

or

$$42982 < \mu < 62002.$$

Solutions To 7-2 Exercises B

7-41 The upper limit of the confidence interval is a z score associated with the degree of confidence. A z score is used here because the sample size exceeds 30. The upper limit is given by

$$77.97 = \bar{x} + E = 76.8 + z(\frac{s}{\sqrt{n}}) = 76.8 + z(\frac{4.2}{5.92}).$$

Solving for z we have

$$.709z = 1.17$$

or $z = 1.649$. A z score of this size corresponds to a 90% confidence interval.

7-43 a. Recognizing that

$$\sqrt{\frac{N - n}{N - 1}} = \frac{\sqrt{N - n}}{\sqrt{N - 1}}$$

we multiply through by the denominator $\sqrt{n}\sqrt{N - 1}$ and get

$$E\sqrt{n}\sqrt{N - 1} = z(\alpha/2)\ \sigma\ \sqrt{N - n}.$$

Squaring both sides we have

$$E^2\ n(N - 1) = [z(\alpha/2)]^2\sigma^2(N - n)$$

$$= [z(\alpha/2)]^2\sigma^2 N - [z(\alpha/2)]^2\sigma^2 n.$$

Putting the two terms involving n on the left hand side we have

$$E^2(N - 1)n + [z(\alpha/2)]^2\sigma^2 n = [z(\alpha/2)]^2\sigma^2 N$$

$$(E^2(N - 1) + [z(\alpha/2)]^2\sigma^2)n = N\sigma^2[z(\alpha/2)]^2.$$

Dividing by the coefficient of n we have

$$n = \frac{N\sigma^2[z(\alpha/2)]^2}{(N - 1)E^2 + \sigma^2[z(\alpha/2)]^2}.$$

b. Using the formula with $z(\alpha/2) = 2.57$ because of the 99% confidence, we have

$$n = \frac{(100)(.000016)(2.57)^2}{99(.000001) + (.000016)(2.57)^2} = \frac{.0106}{.000205} = 52.$$

7-45 Multiplying through by $\sqrt{n}$ gives

$$\sqrt{n}\ E = z(\alpha/2)\sigma.$$

Dividing by E we have

$$\sqrt{n} = \frac{z(\alpha/2)\sigma}{E}.$$

Squaring both sides

$$n = \left[\frac{z(\alpha/2)\sigma}{E}\right]^2.$$

SOLUTIONS TO SECTION 7-3 ESTIMATES AND SAMPLE SIZES OF PROPORTIONS

7-49 a. The best estimate of $\hat{p}$ is $x/n = 100/500 = .20$.

b. The best estimate of $\hat{q}$ is

$$1 - \hat{p} = 400/500 = .80.$$

c. The best point estimate of p is $x/n = 100/500$.

d. The maximum error of the estimate is

$$E = z(\alpha/2)\sqrt{\frac{\hat{p}\hat{q}}{n}} = 1.96\sqrt{\frac{(.20)(.80)}{500}}$$

$$= (1.96)(.0178) = .0351 .$$

7-51 a. The best point estimate of $\hat{p}$ is $325/1068 = .304$.

b. The best point estimate of $\hat{q}$ is $743/1068 = .696$.

c. The best point estimate of p is $\hat{p} = .304$.

d. The maximum error of the estimate is given by

$$E = z(\alpha/2)\sqrt{\frac{\hat{p}\hat{q}}{n}} = 1.96\sqrt{\frac{(.304)(.696)}{1068}}$$

$$= (1.96)(.014) = .0276 .$$

7-53 Our estimate of $\hat{p}$ is $100/400 = .25$ and $\hat{q}$ is .75. Then the maximum error of the estimate is

$$E = z(\alpha/2)\sqrt{\frac{\hat{p}\hat{q}}{n}} = 1.96\sqrt{\frac{(.25)(.75)}{400}} = .042.$$

where $z(\alpha/2) = 1.96$ because of the 95% confidence. Then the 95% confidence interval is given by

$$\hat{p} - E < p < \hat{p} + e$$

or

$$.25 - .042 < p < .25 + .042$$

or

$$.208 < p < .292.$$

7-55 From our 98% confidence $z(\alpha/2) = 2.33$. With $\hat{p} = x/n = (309/512) = .604$, we have $\hat{q} = 1 - \hat{p} = 1 - .604 = .396$. The maximum error of the estimate is then

$$E = z(\alpha/2)\sqrt{\frac{\hat{p}\hat{q}}{n}} = 2.33\sqrt{\frac{(.604)(.396)}{512}} = .05.$$

The 98% confidence interval is then

$$\hat{p} - E < p < \hat{p} + E$$

or

$$.604 - .05 < p < .604 + .05$$

or

$$.554 < p < .654.$$

7-57 Because we wish 90% confidence, then $z(\alpha/2) = 1.64$. Because of the absence of p or q, then the formula is

$$n = \frac{[z(\alpha/2)]^2(.25)}{E^2} = \frac{(1.64)^2(.25)}{(.04)^2} = 421$$

rounding up to the nearest integer.

7-59 From the data the best estimate of $\hat{p}$ is 2307/9261 = .25. Then $\hat{q} = 1 - \hat{p} = 1 - .25 = .75$. The maximum error ofthe estimate for 98% confidence uses z = 2.33 and is

$$E = z(\alpha/2)\sqrt{\frac{\hat{p}\hat{q}}{n}} = 2.33\sqrt{\frac{(.25)(.75)}{9261}} = .010.$$

Then the 98% confidence interval is

$$\hat{p} - E < p < \hat{p} + E$$

or

$$.25 - .01 < p < .25 + .01$$

or

$$.24 < p < .26 .$$

7-61 As our estimate we use a 95% confidence interval. By definition

$\hat{p} = 180/650 = .277$, and $\hat{q} = 1 - p = 1 - .277 = .723$.

For such an interval $z(\alpha/2) = 1.96$ and hence the error in the estimate E is given by

$$E = z(\alpha/2)\sqrt{\frac{\hat{p}\hat{q}}{n}} = 1.96\sqrt{\frac{(.277)(.723)}{650}} = .034.$$

Then the 95% confidence interval is given by

$$\hat{p} - E < p < \hat{p} + E$$

or

$$.277 - .034 < p < .277 + .034$$

or

$$.243 < p < .311.$$

7-63 Using $z(\alpha/2) = 2.57$ because of the 99% confidence, we know that

$\hat{p}$ is given to be .72. Then the error in the estimate is

$$E = z(\alpha/2)\sqrt{\frac{\hat{p}\hat{q}}{n}} = 2.57\sqrt{\frac{(.72)(.28)}{4664}} = .017.$$

Therefore, the 99% confidence interval is given by

$$\hat{p} - E < p < \hat{p} + E$$

or

$$.72 - .017 < p < .72 + .017$$

or

$.703 < p < .737.$

7-65 Since we have an estimate of the size of p, we can use the sample size formula where p is known. Then,

$$n = \frac{[z(\alpha/2)]^2 pq}{E^2} = \frac{(1.645)^2(.85)(.15)}{.025^2} = 553.$$

7-67 Solved like 7-65.

7-69 Solved like 7-61.

7-71 Since we are looking for a 96% confidence interval, the value of z is 2.05. Using p = .81, the maximum error of the estimate is

$$E = z(\alpha/2)\sqrt{\frac{\hat{p}\hat{q}}{n}} = (2.050)\sqrt{\frac{(.81)(.19)}{765}} = .029.$$

The 96% confidence interval will be

$$\hat{p} - E < p < \hat{p} + E$$

or

$$.81 - .029 < p < .81 + .029$$

or

$$.781 < p < .839.$$

7-73 For 99% confidence $z(\alpha/2) = 2.575$. The value of E with

$\hat{p} = 5463/6503 = .84$ and $\hat{q} = .16$ is

$$E = z(\alpha/2)\sqrt{\frac{\hat{p}\hat{q}}{n}} = 2.575\sqrt{\frac{(.84)(.16)}{6503}} = .012.$$

Then the 99% confidence interval is

$$\hat{p} - E < p < \hat{p} + E$$

or

$$.84 - .012 < p < .84 + .012$$

or

$$.828 < p < .852.$$

Since the 99% confidence interval exceeds .80, we are quite confident that the true proportion is more than .80. Therefore, the problem is considered easy.

7-75 As in 7-73 we have the error in the estimate based upon 90% confidence as

$$E = z(\alpha/2)\sqrt{\frac{(.511)(.489)}{5000}} = (1.645)(.007) = .012.$$

Therefore the confidence interval is

$$.511 - .012 < p < .511 + .012$$

or

$$.499 < p < .523.$$

Solutions To 7-3 Exercises B

7-77 A picture of our rectangle would be as shown below. Since 2p + 2q = perimeter = 2, then p + q = 1. We will have maximum area when pq is maximum. This will occur when p = q = .5. This concept corresponds to the ideas discussed when trying to find sample size when p is unknown.

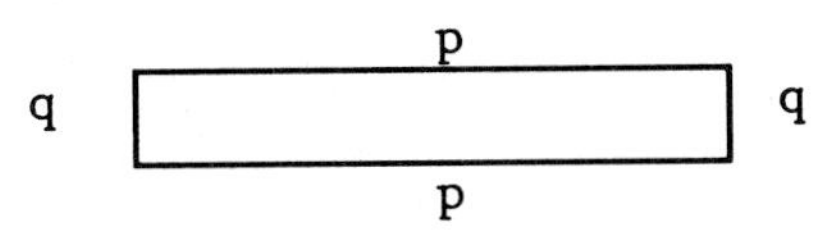

7-79 We can solve the formula for the error in the estimate E for the z score corresponding to the level of confidence. The resultant formula is

$$z(\alpha/2) = E\sqrt{\frac{n}{\hat{p}\,\hat{q}}} = .002\sqrt{\frac{47000}{(.08)(.92)}} = 1.60.$$

This corresponds to a probability of .4452. To get the two tailed probability of a confidence interval, we double this to a probability of .89 or an 89% confidence interval.

SOLUTIONS TO SECTION 7.4 ESTIMATES AND SAMPLE SIZES OF VARIANCES

7-81 a. The best point estimate of σ^2 is $s^2 = 12$.

b. From table 7-1 the sample size is 97.

c. With 25 samples there are 24 degrees of freedom. So the two limits are 9.886 and 45.559 using the .995 and .005 columns.

7-83 a. With 11 scores there will be 10 degrees of freedom. For 95% confidence, we would use the .975 and .025 columns. Then

$$\chi_L^2 = 3.247 \text{ and } \chi_R^2 = 20.483.$$

b. Looking at the 99% confidence table for σ, an error of 5% leads to a sample size of 1335.

c. The best point estimate of σ^2 is s^2 or $(1.2)^2 = 1.44$.

7-85 To construct the 95% confidence interval, we need the two χ^2 values. Using 29 degrees of freedom and the .975 and .025 columns, the two χ^2values are 16.791 and 45.722. Then the 95% confidence interval is

$$\frac{(n-1)s^2}{\chi_R^2} < \sigma^2 < \frac{(n-1)s^2}{\chi_L^2} \text{ or } \frac{29(2.5)^2}{45.722} < \sigma^2 < \frac{29(2.5)^2}{16.791}.$$

Consequently, we have

$$4. < \sigma^2 < 11.3.$$

7-87 With 21 scores we have 20 degrees of freedom. Using the .975 and .025 columns for χ_L^2 and χ_R^2 respectively, we have $\chi_L^2 = 13.844$ and $\chi_R^2 = 41.923$.

The confidence interval is given by

$$\frac{(n - 1)s^2}{\chi_R^2} < \sigma^2 < \frac{(n - 1)s^2}{\chi_L^2}$$

or

$$\frac{(20)(100)}{34.17} < \sigma^2 < \frac{(20)(100)}{9.591}$$

or

$$58.53 < \sigma^2 < 208.52.$$

7-89 Because there are 29 subjects, then there are 28 degrees of freedom. From that row and the .975 and .025 columns, we have

$$\chi_L^2 = 15.308 \text{ and } \chi_R^2 = 44.461.$$

Using the formula for confidence interval on σ^2 the variance we have

$$\frac{(n - 1)s^2}{\chi_R^2} < \sigma^2 < \frac{(n - 1)s^2}{\chi_L^2}$$

or

$$\frac{(28)(17.66)^2}{44.461} < \sigma^2 < \frac{(28)(17.66)^2}{15.308}$$

which becomes

$$196.41 < \sigma^2 < 570.45.$$

Therefore the confidence interval for the standard deviation is

$$14.0 < \sigma < 23.9.$$

7-91 Using 49 degrees of freedom the values of

χ^2_L and χ^2_R are respectively 27.991 and 79.490.

The confidence intervals for standard deviation are

$$\sqrt{\frac{(n - 1)s^2}{\chi^2_R}} < \sigma < \sqrt{\frac{(n - 1)s^2}{\chi^2_L}}$$

or

$$\sqrt{\frac{(49)(9.3)^2}{79.490}} < \sigma < \sqrt{\frac{(49)(9.3)^2}{27.991}} .$$

When we compute the two numbers we get

$$7.30 < \sigma < 12.30.$$

7-93 Since there were 40 students in the sample, there are 39 degrees of freedom. For 99% confidence, the chi square values are given by

$$\chi^2_L = 20.707 \text{ and } \chi^2_R = 66.766.$$

Since $s^2 = 77.6$, our confidence interval would be given by

$$\sqrt{\frac{(n - 1)s^2}{\chi^2_R}} < \sigma < \sqrt{\frac{(n - 1)s^2}{\chi^2}}$$

which becomes

$$\sqrt{\frac{(39)(77.6)}{66.766}} < \sigma < \sqrt{\frac{(39)(77.6)}{20.707}}$$

or

$$6.73 < \sigma^2 < 12.09.$$

7-95 The 100 samples result in 99 degrees of freedom. We have $s^2 = 21.8$. The χ^2 values for 90% confidence use the .95 and .05 columns. The resultant χ^2 values are

$$\chi^2_L = 77.929 \text{ and } \chi^2_R = 124.342.$$

The confidence interval is then

$$\sqrt{\frac{(n - 1)s^2}{\chi^2_R}} < \sigma < \sqrt{\frac{(n - 1)s^2}{\chi^2}}$$

or

$$\sqrt{\frac{(99)(21.8)^2}{124.342}} < \sigma < \sqrt{\frac{(99)(21.8)^2}{77.929}} .$$

or

$$19.45 < \sigma < 24.57.$$

7-97 We determine s^2 from the sample data to be .0136 using the formula of chapter 2. The two values of chi square are

$$\chi_R^2 = 16.75 \text{ and } \chi_L^2 = .412$$

because we have 5 degrees of freedom. The 99% confidence interval is given by

$$\sqrt{\frac{5(.0136)}{16.75}} < \sigma < \sqrt{\frac{5(.0144)}{.412}}$$

or

$$.064 < \sigma < .406.$$

7-99 We find s^2 from the sample as $s^2 = .44$ by using the variance formula of chapter 2. For 9 degrees of freedom and 95% confidence, the left and right values of χ^2 are respectively 2.7 and 19.023. Then the 95% confidence interval is

$$\frac{(n - 1)s^2}{\chi_R^2} < \sigma^2 < \frac{(n - 1)s^2}{\chi_L^2}$$

or

$$\frac{9(.44)}{19.023} < \sigma^2 < \frac{9(.44)}{2.7} \quad \text{or} \quad .21 < \sigma^2 < 1.47.$$

SOLUTIONS TO B EXERCISES

7-101 Using only the upper confidence limit, we have that

$$\frac{(n-1)s^2}{\chi_L^2} = 36 \quad \text{or} \quad \chi_L^2 = \frac{19(3.8)^2}{36} = 7.62.$$

This corresponds to the χ^2 value in the .99 column. This means a confidence level of 98%.

7-103 The 95% confidence interval about σ for large samples uses a z value of 1.96. The form of the interval is

$$\sqrt{\frac{(n - 1)s^2}{\chi_R^2}} < \sigma < \sqrt{\frac{(n - 1)s^2}{\chi_L^2}}$$

The two χ^2 are

$$\chi^2 = \frac{1}{2}[\pm z + \sqrt{2k - 1}]^2 = .5[\pm 1.96 + \sqrt{2(772)-1}]^2$$
$$= .5[\pm 1.96 + 39.28]^2$$
$$= 696.39 \quad \text{and} \quad 850.4.$$

Then the confidence interval is

$$\sqrt{\frac{(771)(2.8)^2}{850.4}} < \sigma < \sqrt{\frac{(771)(2.8)^2}{696.39}}.$$

or

$2.67 < \sigma < 2.94.$

SOLUTIONS TO REVIEW EXERCISES

7-105 a. The best point estimate of μ is $\bar{X} = 83.2$.

b. Since $n > 30$, we use the maximum error of the estimate

$E = z(\alpha/2)(s/\sqrt{n})$.

For 95% confidence $z(\alpha/2) = 1.96$. Then

$$E = 1.96(\frac{4.1}{\sqrt{60}}) = \frac{(1.96)(4.1)}{7.75} = 1.04.$$

Then the 95% confidence interval is

$$\bar{X} - E < \mu < \bar{X} + e$$

or

$$83.2 - 1.04 < \mu < 83.2 + 1.04$$

or

$$82.16 < \mu < 84.24.$$

7-107 We use here the sample size formula for estimating proportions where we have no estimate of p. The value of z for 94% confidence is 1.88. Therefore, the desired sample size is given by

$$n = \frac{z^2(.25)}{E^2} = \frac{1.88^2(.25)}{.05^2} = 354.$$

7-109 We use the sample size formula for estimating the mean. Because of the 95% confidence, the value of the z score is 1.96. Then, the sample size is given by

$$n = \frac{z^2\sigma^2}{E^2} = \frac{(1.96)^2(41)^2}{(4)^2} = 404.$$

7-111 The formula for the confidence interval of a standard deviation is

$$\sqrt{\frac{(n - 1)s^2}{\chi_R^2}} < \sigma < \sqrt{\frac{(n - 1)s^2}{\chi^2}}.$$

The values of χ^2 are based on 99% confidence and 15 df.

$$\chi_L^2 = 4.601 \quad \text{and} \quad \chi_R^2 = 32.801.$$

Then the confidence interval is

$$\sqrt{\frac{(15)(10.5)^2}{32.801}} < \sigma < \sqrt{\frac{(15)(10.5)^2}{4.601}}.$$

or

$$7.1 < \sigma < 19.0.$$

7-113 Since there is 95% confidence the z value is 1.96. Then the error in the estimate is given by

$$E = z(\alpha/2)\sqrt{\frac{\hat{p}\,\hat{q}}{n}} = 1.96\sqrt{\frac{(.17)(.83)}{308}} = .042.$$

Then, the 95% confidence interval is

$$\hat{p} - E < p < \hat{p} + E$$

or

$$.17 - .042 < p < .17 + .042$$

or

$$.128 < p < .212.$$

7-115 a. For $\alpha = .10$, the value of $z(\alpha/2)$ is midway between 1.64 and 1.65, and hence it is 1.645.

b. We have 9 degrees of freedom and use the .975 and .025 columns. Then

$$\chi_L^2 = 2.700 \text{ and } \chi_R^2 = 19.023.$$

c. With 9 degrees of freedom the value will be .05 (two tails) and is 2.262.

d. It occurs when $p = q = .5$. The maximum is .25.

7-117 a. The best point estimate of σ^2 is $s^2 = 16.81$.

b. The two values for χ^2 are based upon the 95% confidence and 15 degrees of freedom. They are

$$\chi_L^2 = 6.262 \quad \text{and} \quad \chi_R^2 = 27.488.$$

Then the confidence interval is

$$\sqrt{\frac{(n-1)s^2}{\chi_R^2}} < \sigma < \sqrt{\frac{(n-1)s^2}{\chi^2}}$$

or

$$\sqrt{\frac{(15)(4.1)^2}{27.488}} < \sigma < \sqrt{\frac{(15)(4.1)^2}{6.262}} .$$

or

$3.03 < \sigma < 6.35.$

7-119 For 99% confidence $z(\alpha/2) = 2.57$. From the data $\hat{p} = .24$. Then $\hat{q} = .76$. The maximum error of the estimate is then

$$E = z(\alpha/2)\sqrt{\frac{\hat{p}\,\hat{q}}{n}} = 2.57\sqrt{\frac{(.24)(.76)}{1998}} = .025.$$

Then the 99% confidence interval for p, the proportion of adults is

$$\hat{p} - E < p < \hat{p} + E$$

or

$$.24 - .025 < p < .24 + .025$$

or

$$.215 < p < .265.$$

7-121 Using the sample size formula for estimating proportions where no estimate $\hat{p}$ is available, we have

$$n = \frac{z(\alpha/2)^2(.25)}{E^2} .$$

Since we are interested in 90% confidence, $z(\alpha/2) = 1.645$. Then

$$n = \frac{(1.645)^2(.25)}{(.05)^2} = 271 \text{ rounding up.}$$

7-123 Although the standard deviation is from the sample, the sample size exceeds 30. Therefore, the formula for the error in the estimate is

$$E = z\left[\frac{s}{\sqrt{n}}\right] = 2.33\,\frac{4}{7.07} = 1.32.$$

Then the confidence interval is given by

$$\bar{x} - E < \mu < \bar{x} + E$$

or

$$16.4 - 1.32 < \mu < 16.4 + 1.32.$$

or

$$15.08 < \mu < 17.72.$$

7-125 The sample size formula for estimating means is

$$n = \left[\frac{z(\alpha/2)\sigma}{E}\right]^2.$$

With 98% confidence, $z(\alpha/2) = 2.33$. Then

$$n = \left[\frac{(2.33)(.8)}{.02}\right]^2 = 8687, \text{ rounding up.}$$

7-127 We standard deviation is from the sample and the sample size is less than 30, so we must use the students t. Because there are 10 subjects, there are 9 degrees of freedom. The students t value is 2.262 for 95% confidence. The error in the estimate is

$$E = t(\alpha/2)\frac{s}{\sqrt{n}} = 2.262$$

Using the computed mean of 6.91, the confidence interval is

$$5.57 < \mu < 8.264.$$

CHAPTER 8. TESTS COMPARING TWO PARAMETERS

SOLUTIONS TO SECTION 8.2 TESTS COMPARING TWO VARIANCES

8-1 Our hypothesis is

$H_0: \sigma^2_A = \sigma^2_B$

$H_1: \sigma^2_A \neq \sigma^2_B \qquad \alpha = .05$

For a test of variances from normal distributions the appropriate test is the F test. We form

$$F = \frac{s^2_A}{s^2_B} = \frac{50}{25} = 2.$$

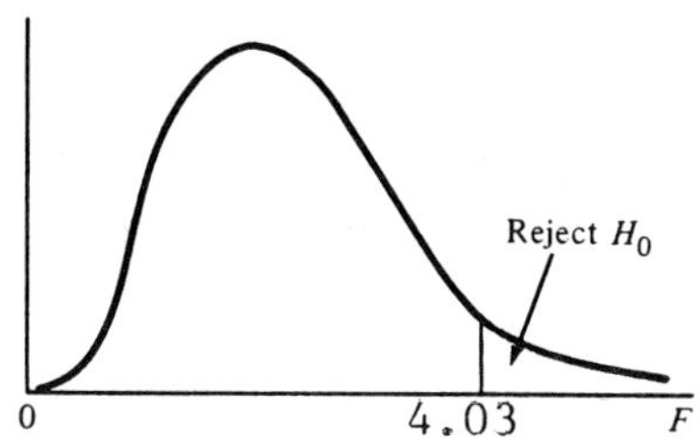

For a two tailed test with $\alpha = .05$, numerator df = 9, and denominator df = 9, the critical F value is 4.026. Our F value of 2 does not exceed this so we fail to reject H_0. The data supports a conclusion of equality of the variances.

8-3 Our hypothesis is as in 8-1.

$H_0: \sigma^2_A = \sigma^2_B$

$H_1: \sigma^2_A \neq \sigma^2_B \qquad \alpha = .05$

Using the F test, we have

$$F = \frac{s^2_A}{s^2_B} = \frac{45.2}{5.8} = 7.79.$$

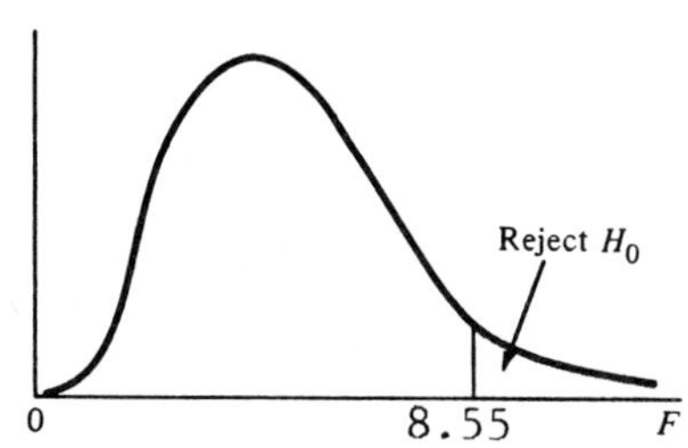

For this two tailed F test with $\alpha = .05$, numerator degrees of freedom = 19 and denominator df = 4, the critical value for F is more than 8.55. Since our observed F does not exceed this, we fail to reject the hypothesis of equality of the variances.

8-5 The null hypothesis is of the form

$H_0: \sigma^2_A = \sigma^2_B$

$H_1: \sigma^2_A \neq \sigma^2_B \qquad \alpha = .05$

Using the F test, we form

$$F = \frac{s^2_B}{s^2_A} = \frac{265}{110} = 2.409.$$

s^2_B is in the numerator because it is the larger variance.

For this two tailed F test with α = .05, the numerator df = 24, and the denominator df is 19. The critical value for F is 2.4523. Since the computed F does not exceed this, we fail to reject the hypothesis of equality of the variances.

8-7 The null hypothesis is

H_0: $\sigma^2_A = \sigma^2_B$

H_1: $\sigma^2_A \neq \sigma^2_B$ α = .05

For our F statistic we compute

$$F = \frac{s^2_A}{s^2_B} = \frac{(14.3)^2}{(1.1)^2} = 169.$$

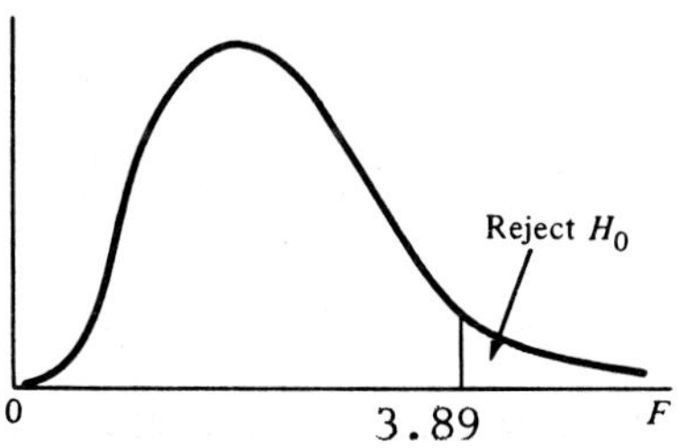

For this two tailed F test, the numerator df is 4 and the denominator df is 14. The critical value for F is 3.8919, since α = .05. The observed value of F exceeds this, consequently, we reject the hypothesis of equality of the two variances.

8-9 Our hypothesis is

H_0: $\sigma^2_A \leq \sigma^2_B$

H_1: $\sigma^2_A > \sigma^2_B$ α = .05

The larger variance is s^2_A so F is given by

$$F = \frac{s^2_A}{s^2_B} = \frac{48}{12} = 4.$$

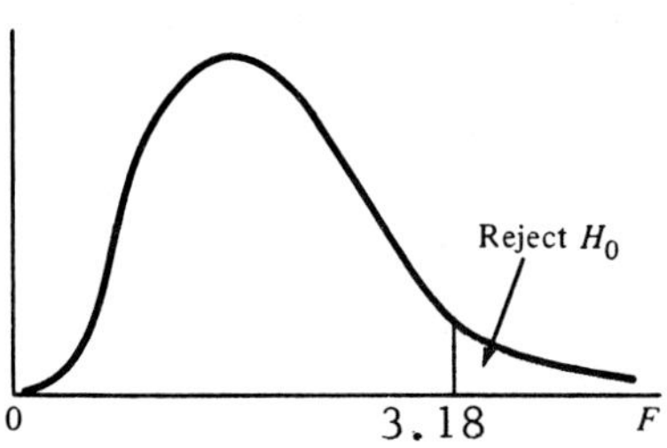

For this one tailed F test, the numerator df is 9 as is the denominator df. Using α = .05, the critical value of F is 3.1789. Since our observed F exceeds this, we reject H_0 and conclude that the data supports the claim that the variance for A is more than the variance for B.

8-11 The null hypothesis is

H_0: $\sigma^2_A \leq \sigma^2_B$

H_1: $\sigma^2_A > \sigma^2_B$ α = .05

Since s^2_A is larger we compute F by

$$F = \frac{s^2_A}{s^2_B} = \frac{225}{160} = 1.41.$$

For this one tailed test with $\alpha = .05$, we have numerator df = 15 and denominator df = 199. As a result the critical value for F is more than 1.6664. Since the computed F does not exceed this, we fail to reject the null hypothesis.

8-13 Our null hypothesis is

$H_0: \sigma^2_M = \sigma^2_W$

$H_1: \sigma^2_M \neq \sigma^2_W \qquad \alpha = .02$

where σ^2_M is the variance for men and σ^2_W is the variance for women. Computing F we have

$$F = \frac{s^2_M}{s^2_W} = \frac{130}{75} = 1.733$$

We use $\alpha = .02$ for this two tailed test. The degrees of freedom for the numerator = 24 and for the denominator = 26. The critical value for F is 2.5848. Since the observed F does not exceed this, we fail to reject the null hypothesis. We conclude that the data does not support a conclusion of inequality of the variances.

8-15 We have data on scale 1 and scale 2 and claim the second produces greater variance. The specific form of the hypothesis is

$H_0: \sigma^2_1 \geq \sigma^2_2$

$H_1: \sigma^2_1 < \sigma^2_2 \qquad \alpha = .05$

We use the F test for the variances

$$F = \frac{s^2_2}{s^2_1} = \frac{(98)^2}{(72)^2} = \frac{9604}{5184} = 1.853.$$

This is a one tailed test with $\alpha = .05$. The degrees of freedom for the numerator is 41-1 = 40. The degrees of freedom for the denominator is 29. The critical value for F is 1.8055. The observed F exceeds this value therefore we reject the null hypothesis. We conclude that the data supports $\sigma^2_1 < \sigma^2_2$.

8-17 Our hypothesis is of the form

$H_0: \sigma_A = \sigma_B$

$H_1: \sigma_A \neq \sigma_B$

where the subscript A denotes those with specific job previews and B for the "no booklet" group. We use the F test as

$$F = \frac{s^2_A}{s^2_B} = \frac{6.56^2}{4.90^2} = 1.79.$$

The numerator degrees of freedom is 59 and the denominator is 39. Using $\alpha = .10$, the critical value is approximately 1.63. Since F exceeds that value, we reject the null hypothesis of equality of variances.

8-19 The hypothesis of equality of the standard deviations takes on the form

$H_0: \sigma_M = \sigma_W$

$H_1: \sigma_M \neq \sigma_W$

where M stands for men and W for women. For the F statistic we have

$$F = \frac{s^2_W}{s^2_M} = \frac{114.16^2}{97.23^2} = 1.38.$$

The numerator df is 67 and the denominator is 85. The critical value for $\alpha = .02$ is more than 1.53. Since the observed value for F is not in the critical region, we fail to reject the null hypothesis of equality of the variances.

8-21 We use the formula from chapter 2 to compute the two variances. For sample A,

$$s^2_A = \frac{n(\sum x^2) - (\sum x)^2}{n(n-1)} = \frac{20(96476) - (1328)^2}{20(19)} = 436.7$$

For sample B

$$s^2_B = \frac{n(\sum x^2) - (\sum x)^2}{n(n-1)} = \frac{16(181168) - (1660)^2}{16(15)} = 596.2$$

Our test is the F test.

H_0: $\sigma^2_A = \sigma^2_B$

H_1: $\sigma^2_A \neq \sigma^2_B$

We compute F as

$$F = \frac{s^2_B}{s^2_A} = \frac{596.2}{436.7} = 1.37$$

This is a two tailed test with $\alpha = .05$. We have 15 df in the numerator and 19 in the denominator. The critical value of F is more than 2.617. Since the computed F will not be more than this,we fail to reject the null hypothesis. The data supports a conclusion of equality of the variances.

8-23 Our hypothesis is

H_0: $\sigma^2_P = \sigma^2_C$

H_1: $\sigma^2_P \neq \sigma^2_C$ $\qquad \alpha = .05$

We use the formula from chapter 2 to compute the sample variances.

$$s^2_P = \frac{n(\sum x^2 f) - (\sum xf)^2}{n(n-1)} = \frac{13(171965) - (1490.5)^2}{13(12)} = 89.5$$

$$s^2_C = \frac{15(202937) - (1740.4)^2}{15(14)} = 71.7.$$

We compute F as 89.5/71.7 = 1.25. For this two tailed test we have $\alpha = .05$. There are 12 df in the numerator and 14 in the denominator. The critical value for F is 3.0502. Since the observed F does not exceed the critical value, we fail to reject the null hypothesis. The data supports a conclusion of equality of the variances.

SOLUTIONS TO 8-2 B EXERCISES

8-25 a. The two degrees of freedom are both 9. From the table the F value is 4.026. The critical value is the reciprocal of that which is .248.

b. Interchanging degrees of freedom, we will have 6 in the numerator and 9 in the denominator. The value for F from Table A-5 is 4.3197. The critical value is the reciprocal of that which is .231.

d. Interchanging the degrees of freedom, we have 9 in the numerator and 24 in the denominator. Using $\alpha = .02$, the value from the F table is 3.256. The critical value is the reciprocal of this which is .307.

8-27 a. Adding a constant to every member of a set of data will not change the variance. Since neither variance is changed, the value of F is unchanged.

b. Both variances will change by the square of the multiplier. Since F is a quotient, the multiplier of the numerator will cancel the multiplier of the denominator. As a result F will remain unchanged.

c. A change to Celsius is accomplished by multiplying each data value by a constant and adding a constant. As a result, as discussed in a. and b., the value of F will remain unchanged.

SOLUTIONS TO SECTION 8-3 TESTS COMPARING TWO MEANS

8-29 We are testing

$H_o: \mu_x = \mu_y$

$H_1: \mu_x \neq \mu_y$

We determine the correct testing method by following the flowchart of figure 8-6 in the text. In this instance σ_x and σ_y are unknown and the sample sizes are over 30. As a result, we use

$$z = \frac{(\bar{x}_C - \bar{x}_E) - (\mu_C - \mu_E)}{\sqrt{\frac{s_C^2}{n_C} + \frac{s_E^2}{n_E}}}$$

$$z = \frac{79.6 - 84.2}{\sqrt{\frac{153.76}{40} + \frac{148.84}{40}}} = -1.67.$$

This is a z statistic with $\alpha = .05$. The critical values for this two tailed test are ± 1.96. Since the test value of z is not in the critical region, we fail to reject the null hypothesis of equality of the two means.

8-31 Our test is

$H_o: \mu_1 = \mu_2$

$H_1: \mu_1 \neq \mu_2$

We follow the flowchart to determine the appropriate test. These are independent samples. From the problem it is apparent that the variances are unknown. The sample sizes are both under 30. Finally, we must test for equality of variances.

$H_0: \sigma^2_1 = \sigma^2_2$

$H_1: \sigma^2_1 \neq \sigma^2_2$

Using the F test on the two variances, we have

$$F = \frac{s^2_1}{s^2_2} = \frac{73.96}{17.64} = 4.19.$$

The critical value from Table A-5 for a two tailed test using $\alpha = .05$ is 2.8621. Since the test value of F exceeds this, we reject the hypothesis of equality of the variances. As a result, the correct t test is as follows.

$$t = \frac{(\bar{x}_1 - \bar{x}_2) - (\mu_1 - \mu_2)}{\sqrt{\frac{s_1^2}{n_1} + \frac{s_2^2}{n_2}}}$$

$$= \frac{98.6 - 97.8}{\sqrt{\frac{73.96}{16} + \frac{17.64}{16}}} = .33$$

The critical t value for this two tailed test using $\alpha = .05$ are ± 2.132. Since the test statistic is not in the critical region, we fail to reject the hypothesis of equality of the two means.

8-33 These are dependent samples because they are data values from the same person in the two positions. We need to generate the differences between data points. If we set values of d as sitting minus supine, the difference data becomes .03, -.64, -.35, .11, .17, .47, .49, .4, .29, .44. Since we are claiming the two positions yield the same mean, then assigning d as sitting-supine, we would be claiming that $\mu_d \neq 0$. The form of our hypothesis is then:

$H_0 : \mu_d = 0$ $\quad H_1 : \mu_d \neq 0$ $\quad \alpha = .05$

The sitting and supine data are both assumed to be normally distributed, and hence the appropriate statistic, because n = the number of data pairs is ≤ 30, is

$$t = \frac{\bar{d} - \mu_d}{s_d/\sqrt{n}}$$. We have $\bar{d} = \frac{1.41}{10} = .141$. Also

$$s_d^2 = \frac{(10)1.35 - .141^2}{90} = .15.$$

Then $t = \frac{.141 - 0}{.122} = 1.16.$

Using 10 - 1 = 9 degrees of freedom and $\alpha = .05$ on a two-tailed test, the critical values of t are -2.262 and 2.262. Since the observed t is not in the critical region, we would conclude that the data supports the hypothesis that $\mu_d = 0$.

8-35 Since this is not dependent data and one of the sample sizes is less than 30, we must first test for equality of the variances

$H_0: \sigma^2_X = \sigma^2_Y$

$H_1: \sigma^2_X \neq \sigma^2_Y$

Using the F test with $\alpha = .05$

$$F = \frac{s^2_Y}{s^2_X} = \frac{324}{196} = 1.65.$$

The critical value of F for α = .05 with 34 degrees of freedom in the numerator and 24 in the denominator is between 2.209 and 2.146. Since the observed value of F will not exceed anything in this interval, we fail to reject the null hypothesis of equality of variances. As a result the correct t test is

$$t = \frac{(\bar{x}_X - \bar{x}_Y) - (\mu_X - \mu_Y)}{\sqrt{\frac{1}{n_X} + \frac{1}{n_Y}}\sqrt{\frac{(n_X - 1)s^2_X + (n_Y - 1)s^2_Y}{n_X + n_Y - 2}}}$$

$$= \frac{(503 - 520)}{\sqrt{\frac{1}{25} + \frac{1}{35}}\sqrt{\frac{24(169)+34(324)}{58}}} = -4.$$

From the t distribution with a two tailed test and α = .05, we have critical values of ±1.96. Since -4. is in the critical region, we reject the null hypothesis of equality of the two means.

8-37 This is not paired data, however the two sample sizes are both greater than 30. Therefore, the form of the test is as follows.

H_o: $\mu_A = \mu_B$

H_1: $\mu_A \neq \mu_B$

where A denotes those given specific job interviews and B those who were not. Then we use a z test as follows.

$$z = \frac{(\bar{x}_A - \bar{x}_B) - (\mu_A - \mu_B)}{\sqrt{\frac{s^2_A}{n_A} + \frac{s^2_B}{n_B}}}$$

$$= \frac{19.14 - 20.81}{\sqrt{\frac{43.03}{60} + \frac{24.01}{40}}} = -1.45.$$

Using α = .10 in this two tailed z test, the critical values are ±1.645. Since the test value of z is not in the critical region, we fail to reject the hypothesis of equality of the two means.

8-39 We are testing

H_o: $\mu_A = \mu_B$

H_1: $\mu_A \neq \mu_B$

The samples are independent and σ_A and σ_B are unknown. As a result we must test for equality of the variances.

$H_0: \sigma^2_A = \sigma^2_B$

$H_1: \sigma^2_A \neq \sigma^2_B$ $\alpha = .05$

We form

$$F = \frac{s^2_A}{s^2_B} = \frac{3.1^2}{1.6^2} = 3.75.$$

This is a two tailed test with numerator degrees of freedom of 11 and denominator df of 13. With $\alpha = .05$ the critical value is less than 3.25. As a result we conclude that the variances are unequal. The test on means uses the following t statistic.

$$t = \frac{(\bar{x}_A - \bar{x}_B) - (\mu_A - \mu_B)}{\sqrt{\frac{s^2_A}{n_A} + \frac{s^2_B}{n_B}}}$$

$$= \frac{14.4 - 16.3}{\sqrt{\frac{9.61}{12} + \frac{2.56}{14}}} = -1.919.$$

This is a two tailed t test and has the smaller of 12 - 1 and 14 - 1 degrees of freedom. Thus df = 11. With $\alpha = .05$ df the critical values for t are ±2.20. The observed t is not in the critical region, therefore we fail to reject the null hypothesis of equality of the means.

8-41 This is paired data since these are before and after treatment data on the same individual. Getting difference data as after-before the data would become

d = 4.1, 4.5, 7.3, 10.7, 6., 7., 6.4, 6.6.

Then $\bar{d} = \frac{52.6}{8} = 6.575$. Also

$$s^2_d = \frac{8(374.36) - 2766.76}{8(7)} = 4.07.$$

The claim is that $\mu_d \neq 0$. Thus

$H_0 : \mu_d = 0$ $H_1 : \mu_d \neq 0.$

Then

$$t = \frac{\bar{d} - \mu_d}{s_d/\sqrt{n}} = \frac{6.575 - 0}{2.02/2.83} = 9.21.$$

A two-tailed t test with $\alpha = .05$ would have critical values of ±2.365 for 7 degrees of freedom. Since the observed t of 9.21 is not between these, we would conclude that the data does support rejection of $\mu_d = 0$ that there is no difference between pain levels before and after hypnosis.

8-43 Since these are two different brands, the two samples are independent. We do not know σ_1^2 and σ_2^2. Since one of the sample sizes is under 30, we must use one of the t test forms. We apply the F test to determine if $\sigma_1^2 = \sigma_2^2$.

$H_0: \sigma^2_1 = \sigma^2_2$

$H_1: \sigma^2_1 \neq \sigma^2_2$ $\alpha = .05$

We form

$$F = \frac{s^2_1}{s^2_2} = \frac{(4.2)^2}{(3.3)^2} = 1.62.$$

The critical value is determined using the F table with $\alpha = .025$ in the right tail. There are 19 degrees of freedom in the numerator and 39 in the denominator. Thus the critical value is larger than 2.0677. The observed value of F is less than this, and hence we have no reason to reject H_0 of equality of variances. The test statistic would then be:

$$t = \frac{(\bar{x}_1 - \bar{x}_2) - (\mu_1 - \mu_2)}{\sqrt{\frac{1}{n_1} + \frac{1}{n_2}}\sqrt{\frac{(n_1 - 1)s_1^2 + (n_2 - 1)s_2^2}{n_1 + n_2 - 2}}}$$

Our hypothesis is for equal nicotine content.

$H_0 : \mu_1 = \mu_2$ $H_1 : \mu_1 \neq \mu_2$.

Then

$$t = \frac{(35.1 - 36.5)}{\sqrt{\frac{1}{20} + \frac{1}{40}}\sqrt{\frac{19(17.64)+39(10.89)}{58}}} = -1.4.$$

There are $n_1 + n_2 - 2 = 58$ degrees of freedom for this two-tailed t test with $\alpha = .05$. The critical t values are $\pm$ 1.96. Since -1.4 is between +1.96 and -1.96, we have no reason to reject H_0 that $\mu_1 = \mu_2$.

8-45 Since this is data on two different groups, the data would be considered independent. The two variances $\sigma_1{}^2$ and $\sigma_2{}^2$ are unknown but the two sample sizes are over 30. As a result, a z test is appropriate.

$$z = \frac{(\bar{x}_M - \bar{x}_F) - (\mu_M - \mu_F)}{\sqrt{\frac{s^2_M}{n_M} + \frac{s^2_F}{n_F}}}$$

The hypothesis under test is that the means are equal. Hence we have

$$z = \frac{525.23 - 538.82}{\sqrt{\frac{9454}{86} + \frac{13032}{68}}} = -.78.$$

The critical value of z would be $\pm$ 2.33 for a two tailed test with α = .02. Since the observed z is inside of that interval, we would fail to reject H_0 in favor of H_1 and conclude that the data supports equality of the two means.

8-47 We are testing $H_0 : \mu_1 \geq \mu_2$ against $H_1 : \mu_1 < \mu_2$. The data is not dependent. Also σ_E and σ_W are unknown. We must run an F test to see if the two variances are the same. Our test is

H_0: $\sigma^2_1 = \sigma^2_2$

H_1: $\sigma^2_1 \neq \sigma^2_2$

The F statistic is

$$F = \frac{s^2_1}{s^2_2} = \frac{5.2^2}{3.2^2} = 2.64.$$

Both the numerator and denominator have 23 degrees of freedom. At the α = .05 level the critical value is less than 2.3567. Since the observed F is in the critical region, we reject the hypothesis of equality of variances. The t test we use is as follows.

$$t = \frac{(\bar{x}_1 - \bar{x}_2) - (\mu_1 - \mu_2)}{\sqrt{\frac{s^2_1}{n_1} + \frac{s^2_2}{n_2}}} = \frac{63 - 60.1}{\sqrt{\frac{(5.2)^2}{24} + \frac{(3.2)^2}{24}}} = 2.32.$$

For a one-tailed t test with α = .05 and 23 df, the critical value is 1.71. Since our t score is in the critical region, we fail to accept H_0. The data supports a conclusion of inequality of the means.

8-49 The variances were tested for equality in exercise 8-23. A conclusion of equality was supported. The two means are 114.7 and 116.0. From this point the test for equality follows the outline of the section.

8-51 This is paired data and so we need to compute the mean and standard deviation of the differences. The individual differences are 36, 50, -15, 2, 0, 22, 36, 6, 26, 13.

$$\bar{d} = \frac{\Sigma d}{n} = \frac{176}{10} = 17.6.$$

$$s_d^2 = \frac{10(6686) - 176^2}{90} = 398.7.$$

Then t is given by

$$t = \frac{\bar{x} - \mu}{s/\sqrt{n}} = \frac{17.6 - 0}{6.3} = 2.79.$$

There are 9 degrees of freedom in this t test. Using α = .05 in this one tailed test, the critical value is t = 1.833. Since the observed t value is in the critical region, we reject the null hypothesis that the mean of the differences is zero. The data supports a conclusion of an increase in the means before and after the course.

8-53 This exercise is solved like 8-51.

8-55 This is not paired dependent data. In addition we do not know $\sigma_1{}^2$ or $\sigma_2{}^2$. The two sample sizes are under 30, and as a result, the test statistic is a t test. We must compute from the data $\bar{x}$ and s for each of the two groups.

$$\bar{x}_M = \frac{368.22}{16} = 23.01 \quad \text{and} \quad s_M = 7.15.$$

$$\bar{x}_F = \frac{207.17}{11} = 18.83 \quad \text{and} \quad s_F = 7.25.$$

The test for equality of the two variances was done in 8-22. The data supports a claim of equality of the variances.

$$t = \frac{(\bar{x}_M - \bar{x}_F) - (\mu_M - \mu_F)}{\sqrt{\frac{1}{n_M} + \frac{1}{n_F}}\sqrt{\frac{(n_M - 1)s_M^2 + (n_F - 1)s_F^2}{n_M + n_F - 2}}}$$

The form of the hypothesis is:

$$H_0 : \mu_M = \mu_F \qquad H_1 : \mu_M \neq \mu_F.$$

Using the data,

$$t = \frac{23.01 - 18.83.}{\sqrt{\frac{1}{16} + \frac{1}{11}}\sqrt{\frac{15(7.15)^2 + 10(7.25)^2}{25}}} = 1.48.$$

The critical values of t based upon $\alpha = .02$, and a two-tailed test with $n_M + n_F - 2 = 25$ degrees of freedom are ± 2.485. Since the observed t does not exceed this, we would fail to reject H_0 in favor of H_1 that the two means differ. The data supports a conclusion of equality of the two means.

8-57 This exercise is solved like 8-55.

8-59 This exercise is solved like 8-52 using means of differences.

SOLUTIONS TO 8-3 B EXERCISES

8-61 In 8-57 we found the differences of the two means to be 12 lb. We have $z(\alpha/2) = 1.96$ from the normal distribution. Also

$$\sqrt{\frac{\sigma^2_1}{n_1} + \frac{\sigma^2_2}{n_2}} = \sqrt{\frac{35^2}{804} + \frac{27^2}{1657}} = 1.40.$$

Then E = (1.96)(1.40) = 2.75. As a result the 95% confidence interval on the difference of the means is
$9.25 = 12 - 2.75 \le \mu_1 - \mu_2 \le 12 + 2.75 = 14.75.$

SOLUTIONS FOR SECTION 8-4 TESTS COMPARING TWO PROPORTIONS

8-65 n_1 = 200 voters in the first sample
n_2 = 400 voters in the second sample
x_1 = 67 as Democrats
x_2 = 148 as Democrats

$$p_1 = \frac{67}{200}, \quad p_2 = \frac{148}{400}.$$

Then $\bar{p} = \dfrac{x_1 + x_2}{n_1 + n_2} = \dfrac{215}{600}$ and $\bar{q} = 1 - \bar{p} = \dfrac{385}{600}$

Then z is given by

$$z = \frac{\frac{x_1}{n_1} - \frac{x_2}{n_2}}{\sqrt{\bar{p}\bar{q}\left[\frac{1}{n_1} + \frac{1}{n_2}\right]}} = \frac{.335 - .37}{\sqrt{(.358)(.642)(.075)}} = -.27.$$

8-67 The two groups contain 1/2 of the total 500 in each and thus $n_1 = 250$ and $n_2 = 250$ subjects. Since 20% of n_1 pass the fitness test, then $x_1 = (.2)(250) = 50$. Similarly $x_2 = (.24)(250) = 60$. Then

$p_1 = .20$ and $p_2 = .24$. Then $\bar{p} = \frac{50 + 60}{500} = \frac{110}{500}$
and $\bar{q} = 1 - \bar{p} = \frac{390}{500}$.
Then z is given by

$$z = \frac{\frac{x_1}{n_1} - \frac{x_2}{n_2}}{\sqrt{\bar{p}\bar{q}\left[\frac{1}{n_1} + \frac{1}{n_2}\right]}} = \frac{.20 - .24}{\sqrt{(.22)(.78)(.008)}} = -1.08.$$

8-69 a. The z statistic is given by the formula:

$$z = \frac{\frac{x_1}{n_1} - \frac{x_2}{n_2}}{\sqrt{\bar{p}\bar{q}\left[\frac{1}{n_1} + \frac{1}{n_2}\right]}} = \frac{\frac{45}{100} - \frac{115}{200}}{\sqrt{(.53)(.47)\left[\frac{1}{100} + \frac{1}{200}\right]}} = -2.08.$$

where $\bar{p} = \frac{x_1 + x_2}{n_1 + n_2} = \frac{160}{300} = .53$.

b. For a two tailed test using normal distributions and $\alpha = .05$, the two critical values are ±1.96.

c. Our form of the test is
$H_0: p_1 = p_2$
$H_1: p_1 \neq p_2$
The z statistic from the data in part a. does not lie between -1.96 and +1.96. Therefore, we would reject the hypothesis of equality of the two proportions.

8-71 a. The z statistic is given by

$$z = \frac{\frac{x_1}{n_1} - \frac{x_2}{n_2}}{\sqrt{\bar{p}\bar{q}\left[\frac{1}{n_1} + \frac{1}{n_2}\right]}} = \frac{\frac{15}{300} - \frac{20}{200}}{\sqrt{.07(.93)(.008)}} = -2.15.$$

b. The critical value for this left tailed test is -2.33.

c. Since the observed t is not less than the critical value, we fail to reject the null hypothesis of equality of the two proportions.

8-73 Our form of the test is
$H_0: p_1 = p_2$
$H_1: p_1 \neq p_2$
From the data $n_1 = (.26)(50) = 13$ and $n_2 = (.20)(50) = 10$. Then $\bar{p} = 23/100 = .23$. The test statistic is

$$z = \frac{\frac{x_1}{n_1} - \frac{x_2}{n_2}}{\sqrt{\bar{p}\bar{q}\left[\frac{1}{n_1} + \frac{1}{n_2}\right]}} = \frac{\frac{13}{50} - \frac{10}{50}}{\sqrt{(.23)(.77)\left[\frac{1}{50} + \frac{1}{50}\right]}} = .71$$

With a two tailed z test using $\alpha = .05$, the critical values are ± 1.96. Since the observed z is not in the critical region, we fail to reject the null hypothesis of equality of the two proportions.

8-75 Our test is of the type
$H_0: p_1 = p_2$
$H_1: p_1 \neq p_2$
From the data on numbers of defects $\bar{p} = (10292 + 26074)$ divided by $(24384 + 60961)$ which is .426. The form of the test is then

$$z = \frac{\frac{x_1}{n_1} - \frac{x_2}{n_2}}{\sqrt{\bar{p}\bar{q}\left[\frac{1}{n_1} + \frac{1}{n_2}\right]}} = \frac{\frac{10292}{24384} - \frac{26074}{60961}}{\sqrt{(.426)(.574)\left[\frac{1}{24384} + \frac{1}{60961}\right]}} = -1.5.$$

The critical values for a two tailed z test with $\alpha = .10$ are ± 1.645. Since the observed value of z is not in the critical region, we fail to reject the null hypothesis of equality of the two proportions.

8-77 This exercise is solved like exercise 8-75.

8-79 The two sample sizes are both 100 students. Also $n_1 = 12$ and $n_2 = 18$. As a result, $\bar{p} = (12 + 18)/(100 + 100) = .15$. The z test is given by

$$z = \frac{\frac{x_1}{n_1} - \frac{x_2}{n_2}}{\sqrt{\bar{p}\bar{q}\left[\frac{1}{n_1} + \frac{1}{n_2}\right]}} = \frac{.12 - .18}{\sqrt{(.15)(.85)(.02)}} = \frac{-.06}{.05} = -1.2.$$

This is a two tailed z test with $\alpha = .05$. The critical values are ± 1.96. Since the test value of z is not in the critical region, we fail to reject the null hypothesis of equality of the variances.

8-81 This exercise is solved as a one tailed test in a manner similar to 8-79.

8-83 Let x_1 denote the number of dropouts from the public institutions and x_2 from the private. Then our hypothesis is that $p_1 = p_2$ with the alternate as $p_1 \neq p_2$. Using $n_1 = 1000$, $x_1 = 300$, $n_2 = 500$, and $x_2 = 130$. Then $\bar{p} = .286$. As a result,

$$z = \frac{\frac{x_1}{n_1} - \frac{x_2}{n_2}}{\sqrt{\bar{p}\bar{q}\left[\frac{1}{n_1} + \frac{1}{n_2}\right]}} = \frac{.3 - .26}{\sqrt{(.286)(.714)(.001 + .002)}} = 1.62.$$

This is a two tailed test with $\alpha = .05$. The critical values are ±1.96. Since the observed value of z is not in the critical region, we fail to reject the null hypothesis of equality of the two proportions.

8-85 This exercise is solved in the same manner as 8-83.

8-87 This exercise is solved in the same manner as 8-83.

SOLUTIONS TO 8-4 B EXERCISES

8-89 There were a total of 250 males who voted of which p_1 the proportion voting Democrat is 128/250 = .512. There were a total of 600 females voting of which 324 voted Democrat. Therefore, the proportion of females voting Democrat is .540. From this point we proceed using the formula of 8-83 to obtain a z value of .745. Since this is a two tailed z test with α = .05, the critical values are ±1.96. The observed z value is not in the critical region, so we fail to reject the null hypothesis of equality of the two proportions.

8-91 Our test hypothesis is
H_0: $p_1 - p_2 = .25$
H_1: $p_1 - p_2 \neq .25$.
Using the test statistic of the formula, we have

$$z = \frac{p_1 - p_2 - .25}{\sqrt{\frac{p_1(1-p_1)}{n_1} + \frac{p_2(1-p_2)}{n_2}}} = \frac{(.42 - .24) - .25}{\sqrt{\frac{(.42)(.58)}{500} + \frac{(.24)(.76)}{500}}} = \frac{-.07}{.03} = -2.33.$$

This is a two tailed z test with $\alpha = .05$. The critical values are ±1.96. Since the observes z of -2.33 is in the critical region, we reject the null hypothesis that the two proportions differ by 25%.

SOLUTIONS TO REVIEW EXERCISES

8-93 The test that both plants have the same rate of defects is a test of the form
H_0: $p_1 = p_2$
H_1: $p_1 \neq p_2$
The test statistic is a z and is computed by

$$z = \frac{\frac{x_1}{n_1} - \frac{x_2}{n_2}}{\sqrt{\bar{p}\bar{q}\left[\frac{1}{n_1} + \frac{1}{n_2}\right]}} = \frac{\frac{6}{400} - \frac{11}{400}}{\sqrt{(.0225)(.9775)\left[\frac{1}{400} + \frac{1}{400}\right]}} = \frac{-.0125}{.01} = -1.25.$$

For this two tailed z test with α = .02 the critical values are ±2.33. Since the test value of z is not in the critical region, we fail to reject the hypothesis of equality of the two proportions.

8-95 The test of equality of standard deviations is the F test of equality of variances. The form of the test is

H_0: $\sigma^2_A = \sigma^2_B$

H_1: $\sigma^2_A \neq \sigma^2_B$ $\alpha = .05$

We form the F value as

$$F = \frac{s^2_B}{s^2_A} = \frac{8.1^2}{6^2} = 1.82.$$

The numerator degrees of freedom is 11 and the denominator is 9. The critical value from the F table is more than 3.86. Since the observed F is not in the critical region, we fail to reject the null hypothesis of equality of the variances and also the standard deviations.

8-97 These are independent samples from two different states. Although σ^2_N and σ^2_A are not known, the two sample sizes exceed 30. As a result the form of the test is

H_0: $\mu_A = \mu_B$

H_1: $\mu_A \neq \mu_B$

We compute the test statistic z by

$$z = \frac{(\bar{x}_A - \bar{x}_B) - (\mu_A - \mu_B)}{\sqrt{\frac{s^2_A}{n_A} + \frac{s^2_B}{n_B}}} = \frac{76 - 64}{\sqrt{\frac{144}{50} + \frac{256}{40}}} = 3.94.$$

With α = .05 for this two tailed test, the critical values are ±1.96. Since the observed z does lie in the critical region, we reject the hypothesis of equality of the two means.

8-99 Because the two carbuerators are on the same cars, the samples are dependent. Setting the differences as B-A, we have differences d as 2.1, 2.1, .5, -.1, -.6, 3.2, .4, 1.5, and 1.8. Then

$$\bar{d} = \frac{10.9}{9} = 1.21 \quad \text{and} \quad s^2_d = \frac{9(25.33) - (10.9)^2}{(9)(8)} = 1.52.$$

and $s/\sqrt{n} = 1.23/\sqrt{9} = .41$.

Our hypothesis is

H_0: $\mu_d = 0$

H_1: $\mu_d \neq 0$.

We use the t test since σ is unknown and n < 30. Thus t = (1.21 - 0)/.41 = 2.95. A two tailed t test with α = .05 and 8 df has critical values of ±2.306. Since the observed t value is in the critical region, we reject the null hypothesis that the means are the same.

8-101 To test the means we first test for equality of the variances because one sample size is under 30. Our test is

H_0: $\sigma^2_A = \sigma^2_B$

H_1: $\sigma^2_A \neq \sigma^2_B$ $\alpha = .05$

We form the test statistic F as

$$F = \frac{s^2_A}{s^2_B} = \frac{2.1^2}{.4^2} = 27.56.$$

The numerator df = 19 and the denominator df = 31. The critical value for F is less than 2.33. Since F is in the critical region, we reject the hypothesis of equality of the variances. For the test on means we have

$H_0: \mu_A = \mu_B$
$H_1: \mu_A \neq \mu_B$

We use the t statistic and compute,

$$t = \frac{(\bar{x}_A - \bar{x}_B) - (\mu_A - \mu_B)}{\sqrt{\dfrac{s^2_A}{n_A} + \dfrac{s^2_B}{n_B}}} = \frac{66.0 - 68.3}{\sqrt{\dfrac{4.41}{20} + \dfrac{.16}{32}}} = \frac{-2.3}{.47} = -4.89.$$

This is a two tailed t test with 20-1=19 degrees of freedom. Using α = .05, the critical values are ±2.093. Since the observed t is in the critical region, we reject the null hypothesis of equality of the two means.

8-103 The form of the test is

$H_0: p_1 = p_2$
$H_1: p_1 \neq p_2$

From the data $x_1 = (.4)(75) = 30$ and $x_2 = (.6)(75) = 45$. We compute the value of z in this test of proportions as

$$z = \frac{\dfrac{x_1}{n_1} - \dfrac{x_2}{n_2}}{\sqrt{\bar{p}\bar{q}\left[\dfrac{1}{n_1} + \dfrac{1}{n_2}\right]}} = \frac{\dfrac{30}{75} - \dfrac{45}{75}}{\sqrt{(.50)(.50)\left[\dfrac{1}{75} + \dfrac{1}{75}\right]}} = \frac{-.2}{.08} = -2.5.$$

This is a two tailed test with α = .10. The critical values are ±1.645. Since the test z is in the critical region, we reject the null hypothesis of equality of the two proportions.

8-105 This is a test of standard deviations. The appropriate test is the F test of variances. The form of the test is

$H_0: \sigma^2_1 = \sigma^2_2$
$H_1: \sigma^2_1 \neq \sigma^2_2$ $\qquad \alpha = .05$

We form the F statistic as

$$F = \frac{s^2_2}{s^2_1} = \frac{78^2}{42^2} = 3.45.$$

There are 11 df for the numerator and 17 for the denominator. The critical value for F is less than 2.93. Since the observed F exceeds the critical value, we reject the hypothesis of equality of the variances.

8-107 We must first test for equality of variances. The results of an F test as done in problem 8-106 for this data leads to a conclusion of no differences in the two variances. Since one of the sample sizes is under 30, we must use a t test for the means. The hypothesis is

$H_0: \mu_A = \mu_B$
$H_1: \mu_A \neq \mu_B$

The appropriate t test is

$$t = \frac{(\bar{x}_A - \bar{x}_B) - (\mu_A - \mu_B)}{\sqrt{\frac{1}{n_A} + \frac{1}{n_B}}\sqrt{\frac{(n_A - 1)s_B^2 + (n_A - 1)s_B^2}{n_A + n_B - 2}}}$$

$$= \frac{43.7 - 48.2}{\sqrt{\frac{1}{40} + \frac{1}{20}}\sqrt{\frac{40(16.2)^2 + 20(16.5)^2}{58}}} = \frac{-4.5}{4.50} = -1.$$

For this t test using $\alpha = .05$ and 58 degrees of freedom, the critical values are approximately ± 1.96. Since the observed t is not on the critical region, we fail to reject the null hypothesis of equality of the two means.

8-109 This is paired data and so we will be using a t test that the differences have a mean of 0. The after distances minus the before distances are in the following list.
620, 60, 105, 190, 475, -40, 160, 220, 160, 370.
These distances have a mean and standard deviation respectively of

$$\bar{x} = \frac{2320}{10} = 232 \quad s^2 = \frac{10(898850) - 2320^2}{90} = 40067$$

Then s = 200.

Our test is of the form
H_0: $\mu \le 0$
H_1: $\mu > 0$.
We form t the test statistic as

$$t = \frac{\bar{x} - \mu}{s/\sqrt{n}} = \frac{232}{200/3.16} = 3.67.$$

There are 10 - 1 = 9 degrees of freedom. For a two tailed test with α = .05, the critical value is 1.833. Since the test value of t is in the critical region, we reject the hypothesis of equality of the difference of the two means to 0. There is a significant difference.

8-111 Because the two sample sizes exceed 30, the form of the test is

H_o: $\mu_X = \mu_Y$

H_1: $\mu_X \ne \mu_Y$

We use the z statistic for the test and form

$$z = \frac{(\bar{x}_X - \bar{x}_Y) - (\mu_X - \mu_Y)}{\sqrt{\frac{s_X^2}{n_X} + \frac{s_Y^2}{n_Y}}}$$

$$z = \frac{22.6 - 31.0}{\sqrt{\frac{33.6}{45} + \frac{22.1}{65}}} = \frac{-8.4}{1.04} = -8.06.$$

Using $\alpha = .025$, the critical values for z are ± 2.24. The observed z is in the critical region and thus we reject the null hypothesis of equality of the two means.

CHAPTER 9. CORRELATION AND REGRESSION

SOLUTIONS TO SECTION 9-2 CORRELATION

9-3 a. For $n = 20$ the critical value is .444. Therefore we reject the null hypothesis of no correlation and conclude that the data supports the existence of positive linear correlation.

b. The critical level is -.444. The value of $r = -.5$ is less than that, so we reject the hypothesis of no correlation. The data supports a conclusion of negative linear correlation.

c. For $n = 50$ the critical value is .279. The value of $r = .2$ is less than that level so we fail to reject the null hypothesis of no correlation.

9-5 a. The scatter diagram is as given below.

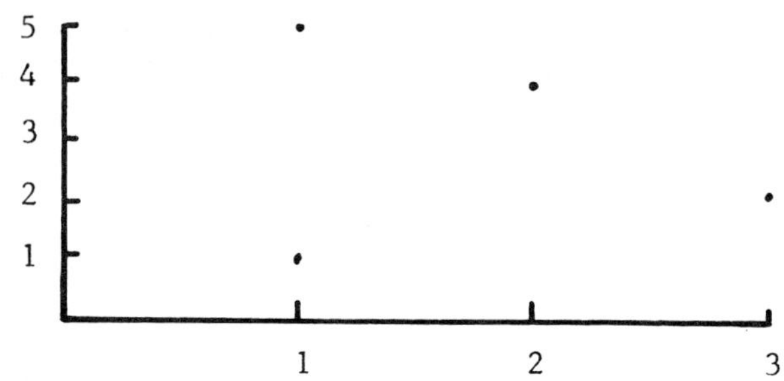

b. n = the number of pairs = 4
c. $\sum x = 1 + 1 + 2 + 3 = 7$
d. $\sum x^2 = 1 + 1 + 4 + 9 = 15$
e. $(\sum x)^2 = (7)^2 = 49$
f. $\sum xy = 1x1 + 1x5 + 2x4 + 3x2 = 20$
g. By formula 9-1

$$r = \frac{n(\sum xy) - (\sum x)(\sum y)}{\sqrt{n(\sum x^2)-(\sum x)^2}\ \sqrt{n(\sum y^2)-(\sum y)^2}}$$
$$= \frac{4(20) - (7)(12)}{\sqrt{4(15)-(49)}\ \sqrt{4(46)-(144)}} = -.191.$$

9-7 a. The scatter diagram is as follows.

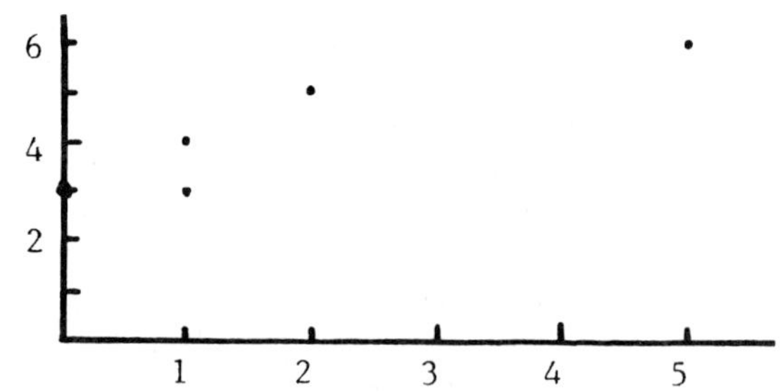

b. There are 5 pairs of points so $n = 5$.
c. $\sum x = 0 + 1 + 1 + 2 + 5 = 9$
d. $\sum x^2 = 31$
e. $(\sum x)^2 = 9^2 = 81$
f. $\sum xy = 3 + 4 + 10 + 30 = 47$
g. The correlation coefficient is given by

$$r = \frac{n(\sum xy) - (\sum x)(\sum y)}{\sqrt{n(\sum x^2)-(\sum x)^2}\ \sqrt{n(\sum y^2)-(\sum y)^2}}$$
$$= \frac{5(47)-9(21)}{\sqrt{5(31)-81}\ \sqrt{5(95)-441}} = .917.$$

9-9 a. Let x be the number of drinks and y be the blood alcohol concentration. Then the scatter diagram is as follows.

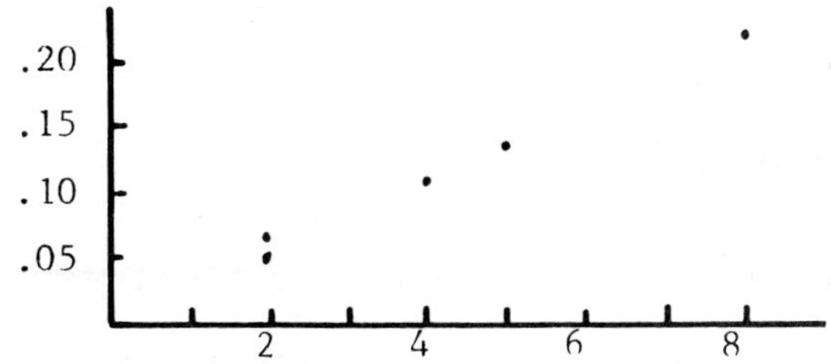

b. We compute r based on formula 9.1.

$$r = \frac{n(\sum xy) - (\sum x)(\sum y)}{\sqrt{n(\sum x^2)-(\sum x)^2}\,\sqrt{n(\sum y^2)-(\sum y)^2}} = \frac{5(3.07) - 21(.57)}{\sqrt{5(113)-(21)^2}\,\sqrt{5(.0835)-(.57)^2}} = .997.$$

c. For n = 5 the critical value is .878 for an $\alpha = .05$

d. The observed correlation is more than the critical value, so we reject the null hypothesis. The data supports a conclusion of positive linear correlation between numbers of drinks and blood alcohol concentration.

9-11 a. The scatter diagram using x as exports and y as imports is as follows.

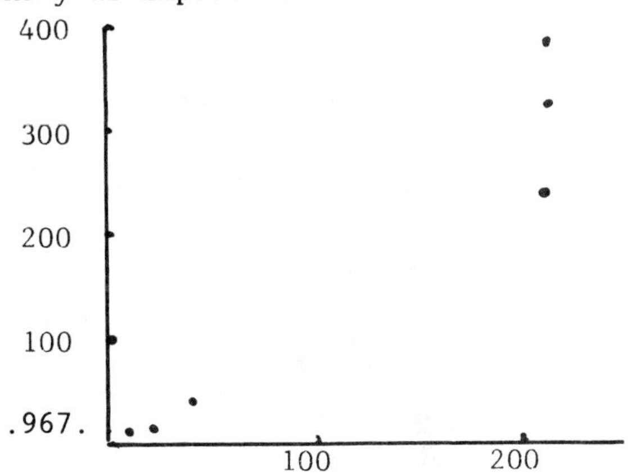

b. Computing r by formula 9.1 we have

$$r = \frac{n(\sum xy) - (\sum x)(\sum y)}{\sqrt{n(\sum x^2)-(\sum x)^2}\,\sqrt{n(\sum y^2)-(\sum y)^2}} = \frac{6(207983) - (730)(1005)}{\sqrt{6(146238)-(730)^2}\,\sqrt{6(305107)-(1005)^2}} = .967.$$

c. Using $\alpha = .05$, the critical value for r with n = 6 is .811.

d. Since the observed r exceeds the critical value, we reject the null hypothesis and conclude that the data supports a conclusion of significant positive linear correlation.

9-13 a. For x as beer consumption and y as wine consumption, the scatter diagram is as given below.

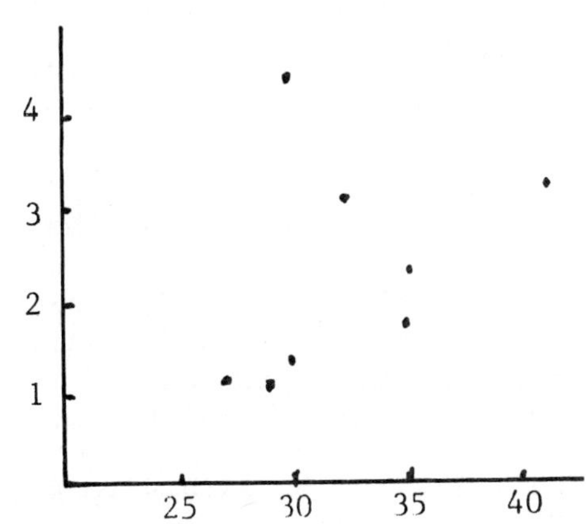

b. By using formula 9.1, r is

$$r = \frac{n(\sum xy) - (\sum x)(\sum y)}{\sqrt{n(\sum x^2)-(\sum x)^2}\,\sqrt{n(\sum y^2)-(\sum y)^2}}$$

$$= \frac{8(602.36)-(258.6)(18.3)}{\sqrt{8(8515)-(258.6)^2}\ \sqrt{8(50.99)-(18.3)^2}} = .287.$$

c. For 8 pairs of observations and $\alpha = .05$ the critical value of r is .707.

d. Since the observed r does not exceed the critical value, we fail to reject the null hypothesis of no correlation.

9-15 a. With x as verbal score and y as math score, the scatter diagram is as follows.

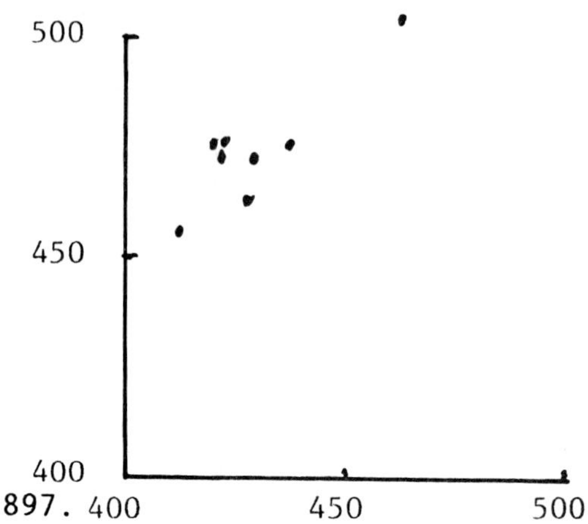

b. We compute r as

$$r = \frac{n(\sum xy) - (\sum x)(\sum y)}{\sqrt{n(\sum x^2)-(\sum x)^2}\ \sqrt{n(\sum y^2)-(\sum y)^2}}$$

$$= \frac{8(1626081) - (3437)(3781)}{\sqrt{8(1478087)-(3437)^2}\ \sqrt{8(1789357)-(3781)^2}} = .897.$$

c. For 8 samples and $\alpha = .05$, the critical value is .707.

d. Since the observed correlation coefficient exceeds the critical value, we reject the null hypothesis of no correlation. The data supports a conclusion of positive linear correlation between verbal and math SAT scores.

9-17 a. Using x as automatic weapons and y as murder rate, the scatter diagram is as follows.

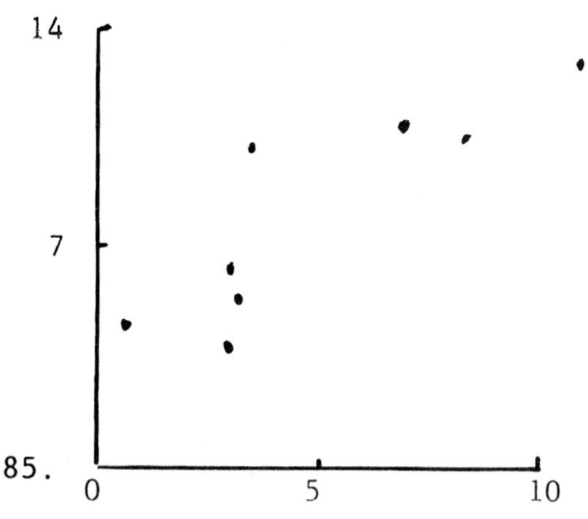

b. Using formula 9.1 we compute r.

$$r = \frac{n(\sum xy) - (\sum x)(\sum y)}{\sqrt{n(\sum x^2)-(\sum x)^2}\ \sqrt{n(\sum y^2)-(\sum y)^2}}$$

$$= \frac{8(397.2)-(38.5)(65.2)}{\sqrt{8(283.2)-(38.5)^2}\ \sqrt{8(622.2)-(65.2)^2}} = .885.$$

c. For $\alpha = .05$ and $n = 8$, the critical value for r is .707.

d. Since the observed correlation coefficient exceeds the critical value, we reject the null hypothesis of no correlation. We conclude that the data supports the position of positive linear correlation between registered automatic weapons and the murder rate.

9-19 This exercise is solved in a manner similar to 9-17.

9-21 This exercise is solved in a manner similar to 9-17.

9-23 a. Using x as the tax rate and y as the selling price, the scatter diagram is as follows.

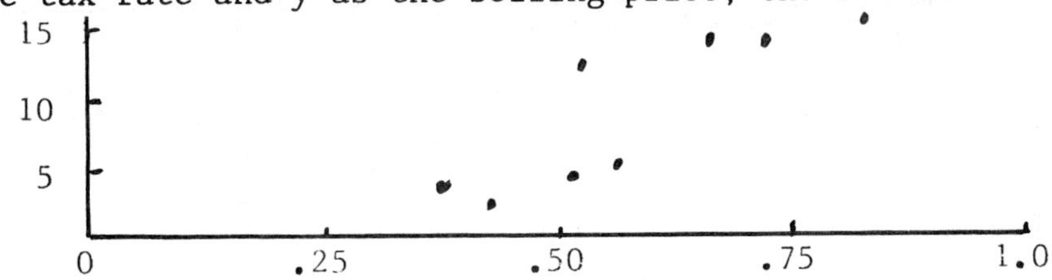

b. The correlation coefficient is computed from the formula.

$$r = \frac{n(\sum xy) - (\sum x)(\sum y)}{\sqrt{n(\sum x^2)-(\sum x)^2}\ \sqrt{n(\sum y^2)-(\sum y)^2}}$$

$$= \frac{8(3197.5)-(17.4)(1334)}{\sqrt{8(43.78)-(17.4)^2}\ \sqrt{8(241394)-(1334)^2}} = .883.$$

c. The critical value from the table using $\doteq$ = .05 and n = 8 is .707.

d. Since the observed value of r exceeds the critical value, we reject the null hypothesis of no correlation. The data supports a conclusion of positive linear correlation between the variables.

9-25 a. With x as productivity and y as dexterity, the scatter diagram is given by

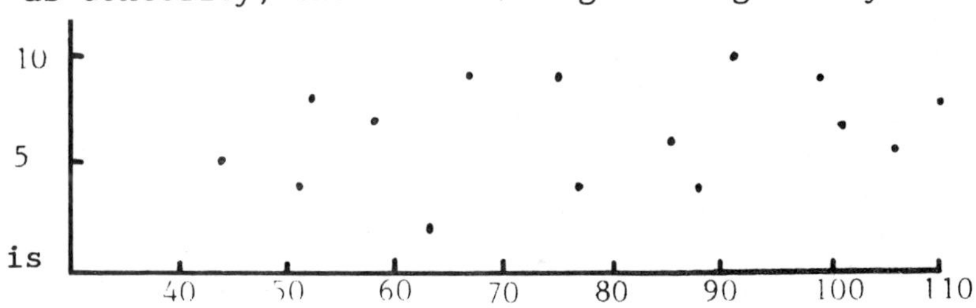

b. The correlation coefficient is

$$r = \frac{n(\sum xy) - (\sum x)(\sum y)}{\sqrt{n(\sum x^2)-(\sum x)^2}\ \sqrt{n(\sum y^2)-(\sum y)^2}}$$

$$= \frac{15(7850)-(1168)(98)}{\sqrt{15(97376)-(1168)^2}\ \sqrt{15(718)-(98)^2}} = .310.$$

c. With α = .05 and n = 15 samples, the critical correlation coefficient is .514.

d. Since the observed correlation coefficient does not exceed the critical value, we fail to reject the null hypothesis of no correlation. The data supports a conclusion of no correlation between productivity and dexterity.

9-27 Solved in a manner similar to 9-25.

SOLUTIONS TO 9-2 B EXERCISES

9-29 There will be respectively 2, 3, 8, 18, 28, 38, 48, and 98 degrees of freedom. Then for α = .10 the corresponding values of students t are 2.92, 2.353, 1.86, 1.734, 1.701, and the others are 1.645. Using the formula, the corresponding values of r are .9, .805, .549, .378, .306, .258, .231, and .164.

9-31 The scatter diagram is as follows.

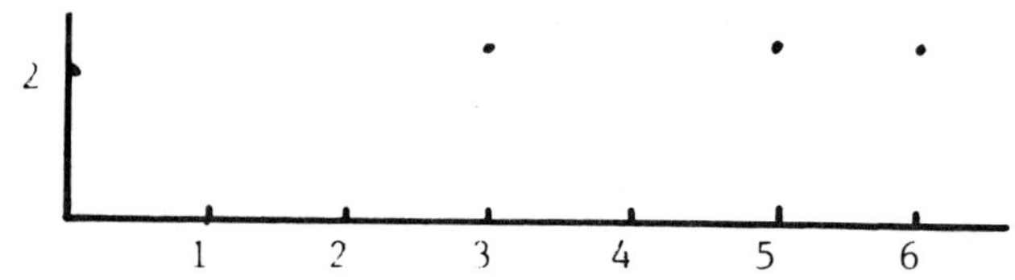

We have the following results computed from the data.
$n = 5$
$\sum x = 19$
$\sum y = 10$
$\sum xy = 38$
$\sum x^2 = 95$
$\sum y^2 = 20$.
To compute r the numerator will be $n\sum xy - \sum x \sum y = 5(38)-(19)(10) = 0$.
It is also the case that the denominator is 0 because for that part associated with y we have $n\sum y^2 - (\sum y)^2 = 5(20)-100 = 0$. This is a degenerate situation in which we can perfectly establish a value of $y = 2$ for any given x, and hence we have perfect correlation. On the other hand it is impossible to compute r using the formula. Since all values of y are 2 as x varies, it is apparent that x and y are unrelated. There is no correlation.

9-33 From problem 9-10, the correlation was .997. Changing the last length from 12 to 36 will change the correlation coefficient. Using the formula of the section, the new correlation coefficient will be .912. Extreme values in data can affect correlations to a significant extent in some cases.

SOLUTIONS TO SECTION 9-3 REGRESSION

9-37 The scatter diagram and regression line are plotted on the graph below.

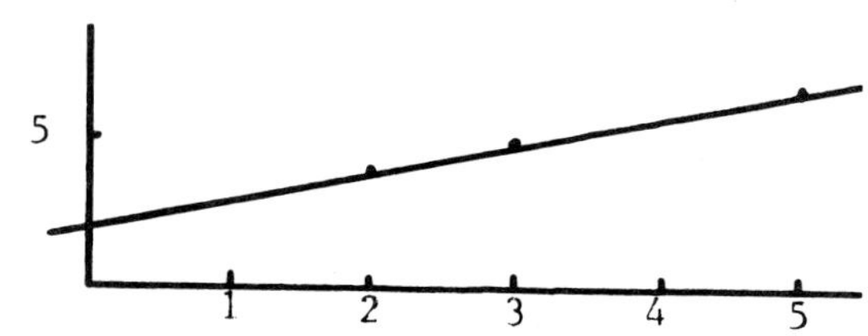

We use formula 9-2 to compute the slope.

$$m = \frac{n\sum xy - \sum x\sum y}{n(\sum x^2)-(\sum x)^2} = \frac{3(58) - (10)(16)}{3(38)-(10)^2} = 1.$$

Also $\bar{x} = \frac{10}{3}$, $\bar{y} = \frac{16}{3}$. Then $b = \bar{y} - m\bar{x} = \frac{16}{3} - \frac{10}{2} = 2$.

As a result the equation is $y = x + 2$.

9-39 We use the values generated in problem 9-5 for $\sum x$, $\sum y$, $\sum x^2$, and $\sum xy$. Then the slope of the regression line is

$$m = \frac{n\sum xy - \sum x\sum y}{n(\sum x^2)-(\sum x)^2} = \frac{4(20)-(7)(12)}{4(15)-(7)^2} = -.36.$$

We use the equation of the line itself to compute the value of the y intercept, b.
$\bar{x} = \frac{7}{4}$, and $\bar{y} = \frac{12}{4} = 3$. Then $b = \bar{y} - m\bar{x} = 3 - (-.36)(1.75) = 3.63$.

Then the equation of the regression line is
$y = -.36x + 3.63$.

9-41 Using the data of problem 9-9, we compute the slope of the regression line as

$$m = \frac{n\sum xy - \sum x\sum y}{n(\sum x^2)-(\sum x)^2} = \frac{5(3.07)-(21)(.57)}{5(113)-(21)^2} = .027.$$

We use the equation of the line to compute the y intercept.

$\bar{x} = \frac{21}{5} = 4.2$ and $\bar{y} = \frac{.57}{5} = .114$. Then $b = .114 -(.027)(4.2) = .0006$.

Then the equation of the regression line is

$$y = .027x + .0006.$$

9-43 Using the data from 9-11 to compute the slope of the regression line, we have

$$m = \frac{n\sum xy - \sum x\sum y}{n(\sum x^2)-(\sum x)^2} = \frac{6(207983) - (730)(1005)}{6(146238)-(730)^2} = 1.49.$$

From the fact that $b = \bar{y} - m\bar{x}$ we need to compute

$$\bar{x} = \frac{730}{6} = 121.67 \quad \text{and } \bar{y} = \frac{1005}{6} = 167.5.$$

Then $b = 167.5 - (1.49)(121.67) = -14.1$. The line is then

$$y = 1.49x - 14.1.$$

9-45 We use the data from problem 9-13 to compute the slope of the regression line.

$$m = \frac{n\sum xy - \sum x\sum y}{n(\sum x^2)-(\sum x)^2} = \frac{8(602.4)-(258.6)(18.3)}{8(8515.2)-(258.6)^2} = .0693.$$

Since $b = \bar{y} - m\bar{x}$, we need to compute $\bar{x} = \frac{258.6}{8} = 32.3$ and $\bar{y} = \frac{18.3}{8} = 2.3$. Then b is given by $b = 2.3 - (32.3)(.069) = .0464$. As a result the equation of the regression line is

$$y = mx + b = .069x + .0464$$

9-47 This is also the data from problem 9-15. We use it to evaluate the slope of the regression line.

$$m = \frac{n\sum xy - \sum x\sum y}{n(\sum x^2)-(\sum x)^2} = \frac{8(1626081) - (3437)(3781)}{8(1478087) - (3437)^2} = 1.14.$$

We have $\bar{x} = \frac{3437}{8} = 429.6$ and $\bar{y} = \frac{3781}{8} = 472.6$.

Then $b = 472.6 - (1.14)(429.6) = -16.5$.

As a result the equation of the regression line is

$$y = 1.14x - 16.5.$$

9-49 This uses the same data as problem 9-17. We use that data to evaluate the slope of the regression line.

$$m = \frac{n\sum xy - \sum x\sum y}{n(\sum x^2)-(\sum x)^2} = \frac{8(397.2) - (38.5)(65.2)}{8(283.15)-(38.5)^2} = .85.$$

We have $\bar{x} = \frac{38.5}{8} = 4.81$ and $\bar{y} = \frac{65.2}{8} = 8.15$.

Computing the y intercept we have $b = \bar{y} - m\bar{x}$.

Then $b = 8.15 - (4.81)(.85) = 4.05$. Then the equation of the regression line is $y = .85x - 4.05$.

9-51 This problem is solved in a manner similar to 9-49.

9-53 This exercise is solved in the same manner as 9-49.

9-55 This data is the same as that of problem 9-23. We compute the slope of the regression line

$$m = \frac{n\sum xy - \sum x \sum y}{n(\sum x^2) - (\sum x)^2} = \frac{8(3197.5) - (17.4)(1334)}{8(43.78) - (17.4)^2} = 49.88.$$

Also $\bar{x} = \frac{17.8}{8} = 2.225$ and $\bar{y} = \frac{1334}{8} = 166.75$.

The y intercept is given by $b = \bar{y} - m\bar{x} = 166.75 - (49.88)(2.225) = 58.26$.

The equation of the regression line is

$$y = 49.88x + 58.26.$$

9-57 a. Taking x as productivity and y as dexterity, the scatter diagram and sketch of the line are given below.

b.

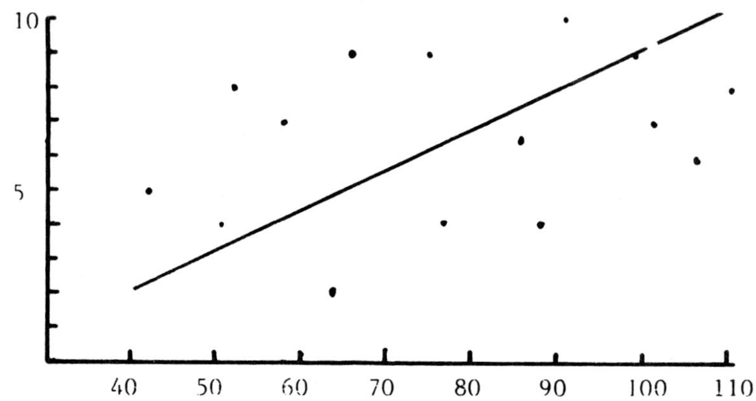

c. Using formula 9-2 to find the slope, we have

$$m = \frac{n\sum xy - \sum x \sum y}{n(\sum x^2) - (\sum x)^2} = \frac{15(7850) - (1168)(98)}{15(97376) - (1168)^2} = \frac{3268}{96416} = .034.$$

Also $\bar{x} = \frac{1168}{15} = 77.87$ and $\bar{y} = \frac{98}{15} = 6.53$. Then b is given by

formula 9-4 as $b = \bar{y} - m\bar{x} = 6.53 - (.034)(77.87) = 3.88$.

Then the regression line equation as computed from the data is

$$y = .034x + 3.88.$$

9-59 a. With x as the cigarette consumption and y the percentage of psychiatric admissions, the scatter diagram and sketch are given below.

b.

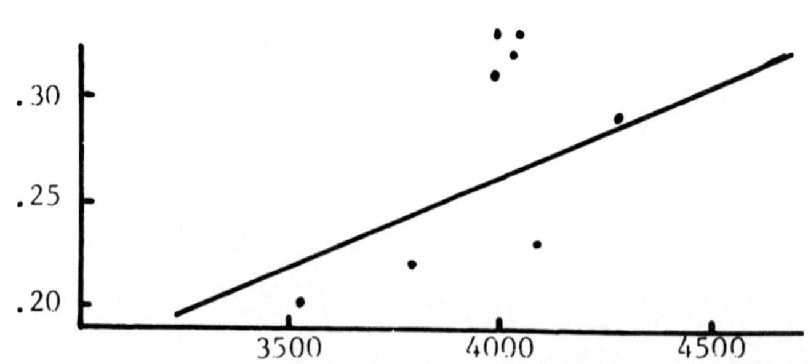

c. With x and y as defined in a above, the correlation coefficient is given by

$$m = \frac{n(\sum xy) - (\sum x)(\sum y)}{n(\sum x^2) - (\sum x)^2}$$

$$= \frac{8(8868.97) - (31607)(2.23)}{8(125348161)-(31607)^2} = \frac{468.15}{3593500} = .00013.$$

For formula 9 we need $\bar{x} = \frac{31607}{8} = 3951$ and $\bar{y} = \frac{2.23}{8} = .279$.

Then b = .279 - (.00013)(3951) = -.235. Then the equation for the regression line is

y = .00013x - .235

9-61 a. For n = 100 the critical value of r is .196 for α = .05. Since the observed r of .999 exceeds the critical value, we accept the conclusion that the data is correlated. Therefore, the best estimate of y is from the regression line which is

y' = (50)(2.) + 10 = 110.

b. For n = 10 and α = .05, the critical r is .632. Since the observed r of .005 does not exceed this, the best estimate of y is the mean $\bar{y}$ = 25.

c., d., and e. are solved in the same way as a and b.

9-63 a. For n = 40 and α = .05 the critical value for a test of a null hypothesis is .312. Since the observed r does not exceed this, we fail to reject the hypothesis of no correlation As a result the best estimate of y is its overall mean 6.

b. Here we have the same critical value as in a. Hence, the observed value of r is more than the critical value, and we feel that there is positive linear correlation between the two variables. Consequently we use the regression equation for our estimation. The predicted value for y is (3)(5) + 2 = 17.

c, d, and e are solved in a manner similar to a and b.

SOLUTIONS TO 9-3 B EXERCISES

9-65 The revised data is as follows.

Exports	10	20	43	221	2180	2180
Imports	9	15	40	245	326	370

For x as exports and y as imports we compute the slope of the regression line as follows.

$$m = \frac{n\sum xy - \sum x\sum y}{n(\sum x^2)-(\sum x)^2} = \frac{6(847595)-(2692)(1005)}{6(4851114)(2692)^2} = .133.$$

Then $\bar{x} = \frac{2693}{6} = 448.83$ and $\bar{y} = \frac{1005}{6} = 167.5$.

Then b = 167.5 - (.133)(448.8) = 108. The regression equation is

y = .133x + 108.

This is a substantial change from the regression equation of problem 9-43.

9-67 From formula 9-5 we have

$r = \frac{ms_x}{s_y}$. Therefore, $rs_y = ms_x$.

If r = 1/2 and m = 1/2, then $s_y = s_x$.

9-69 Upon looking at the formulas for r and m, we note that the two numerators are the same. Since the denominators for both are always positive, the sign for both r and m will be the sign of that numerator.

SOLUTIONS TO SECTION 9-4 VARIATION

9-73 Given r = .333, the coefficient of determination is r^2 = .111. That is, 11.1% of the total variation can be explained by the regression line.

9-75 Since r^2 = .64, then 64% is explained.

9-77 We use a table as in the section. The mean $\bar{y}$ = 61/5 = 12.2.

x	y	y'	$y-\bar{y}$	$(y-\bar{y})^2$	$y'-\bar{y}$	$(y'-\bar{y})^2$
1	5	5	-7.2	51.84	-7.2	51.84
2	8	8	-4.2	17.64	-4.2	17.64
3	11	11	-1.2	1.44	-1.2	1.44
5	17	17	4.8	23.04	4.8	23.04
6	20	20	7.8	60.84	7.8	60.84
	Totals			154.8		154.8

We can now proceed with the questions.

a. The explained variation is the sum in the last column which is 154.8.

b. The unexplained variation is the difference of the two totals which is 0.

c. The total variation is the sum of $(y-\bar{y})^2$ = 154.8.

d. The coefficient of determination is r^2 which is the explained variation over the total variation. In this case that is 1.

e. For each value of x, y and y' are the same. Therefore, y-y' = 0. As a result, $\sum(y-y')^2$ = 0. Thus the standard error of the estimate is 0.

9-79 We generate the table as in 9-77 above. Computing, we have $\bar{y}$ = 13.4.

x	y	y'	$y-\bar{y}$	$(y-\bar{y})^2$	$y'-\bar{y}$	$(y'-\bar{y})^2$
4	13	12.78	-.4	.16	.62	.38
7	22	22.03	8.6	73.96	8.63	74.48
5	16	15.86	2.6	6.76	2.46	6.05
2	7	6.62	-6.4	40.96	6.78	45.97
3	9	9.7	-4.4	19.36	3.7	13.69
	Totals			141.2		140.57

We can from this answer the questions a-e.

a. The total explained variation is the sum in the final column which is 140.57.

b. The unexplained variation is the difference of the two sums which is 141.2 - 140.57 = .63.

c. The total variation is the sum 141.2.

d. The coefficient of determination is the quotient of the explained variation over the total variation. Here it is 140.57/141.2 = .9955.

e. The values of y-y' are respectively .22, .03, .14, .38, and .7. The sum of the squares of these values is .7. Then the standard error of the estimate is

$$s_e = \sqrt{\frac{\Sigma(y-y')^2}{n-2}} = \sqrt{\frac{.7}{3}} = .48.$$

9-81 a. The regression equation is y = 3x + 2. For x = 4, y' = 14.

b. Since $s_e = 0$, there is no error in our estimate of y.

9-83 a. For x = 6, y = 3.0811(6) + .45946 = 18.94.

b. The standard error of the estimate is .48. We compute

$$E = s_e t(\alpha/2) \sqrt{1 + \frac{1}{n} + \frac{n(x_0 - \bar{x})^2}{n(\sum x^2) - (\sum x)^2}}\ .$$

We have $x_0 = 6$, $\bar{x} = \frac{21}{5} = 4.2$.

$\sum x = 21$ and $\sum x^2 = 16 + 49 + 25 + 4 + 9 = 103$.

We are using $\alpha = .05$. There are 5 - 2 = 3 degrees of freedom. Therefore,

$$t(\alpha/2) = 3.182.$$

Therefore,

$$E = (.48)(3.182) \sqrt{1 + \frac{1}{5} + \frac{5(6 - 4.2)^2}{5(103) - (21)^2}} = 1.82.$$

Then the 95% confidence interval is

$$y' - E < \mu < y' + E$$

$$18.94 - 1.82 < \mu < 18.94 + 1.82$$

or

$$17.12 < \mu < 20.76.$$

SOLUTIONS TO 9-4 B EXERCISES

9-85 a. Since the unexplained variation is $\sum(y-y')^2$, we have

$$s_e = \sqrt{\frac{\text{unexplained variation}}{n-2}}$$

or

$$\text{unexplained variation} = s_e^2(n-2)$$

b. Let e be the explained variation. Then,

$$r^2 = \frac{e}{\text{total variation}} = \frac{e}{e + \text{unexplained variation}}\ .$$

Then solving this for e we have

$$e = \frac{r^2(\text{unexplained variation})}{1 - r^2}.$$

c. r = the square root of r^2 = -.949. We select the negative square root because the regression line has a negative slope.

SOLUTIONS TO 9-5. MULTIPLE REGRESSION

9-89 The predicted value is

$$y' = 34.8 + (1.21)24 + .23(92) = 85.$$

9-91 The predicted value is

$$y' = 34.8 + (1.21)18 + (.23)81 = 75.2.$$

9-93 From the coefficient column,

$$y' = -127.7 + 1.034x_1 + 12.434x_2 + 2.296x_3 - 1.059x_4.$$

9-95 Substituting into the equation of 9-93, we have

$$y' = -127.7 + (1.034)7 + (12.434)20 + (2.296)75 - (1.059)42$$
$$= 255.94.$$

Exercises 9-97 through 9-105 are solved using a computer program.

SOLUTIONS TO REVIEW EXERCISES

9-111 a. By the formula for r with x as midterm grade and y as final grade, we have

$$r = \frac{n(\sum xy) - (\sum x)(\sum y)}{\sqrt{n(\sum x^2)-(\sum x)^2}\ \sqrt{n(\sum y^2)-(\sum y)^2}}$$
$$= \frac{5(34265)-(390)(435)}{\sqrt{5(30898)-(390)^2}\ \sqrt{5(38123)-(435)^2}} = \frac{1675}{(48.9)(37.3)} = .918.$$

b. From table A-7 the critical value for r using α = .05 and n = 5 is .878.

c. Since the observed r exceeds the critical value, we conclude that the data supports positive linear correlation between the variables of midterm grade and final exam grade.

d. The slope of the regression line is given by

$$m = \frac{n\sum xy - \sum x\sum y}{n(\sum x^2)-(\sum x)^2} = \frac{5(34265)-(390)(435)}{5(30898)-(390)^2} = .701.$$

The two means are $\bar{x} = \frac{390}{5} = 78$ and $\bar{y} = \frac{435}{5} = 87$.

Then the y intercept is given by $\bar{y} - m\bar{x}$ = 32.32.

The equation of the regression line is given by

$y = .701x + 32.32.$

e. The scatter diagram and regression line are graphed below.

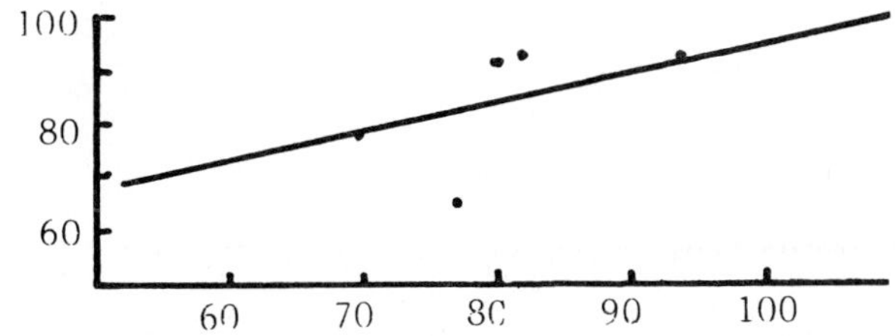

9-113 a. With x as motivation test score and y as personality test score, we compute the correlation coefficient as

$$r = \frac{n(\sum xy) - (\sum x)(\sum y)}{\sqrt{n(\sum x^2) - (\sum x)^2}\sqrt{n(\sum y^2) - (\sum y)^2}}$$

$$= \frac{6(4775) - (290)(97)}{\sqrt{6(14406) - (290)^2}\sqrt{6(1609) - (97)^2}} = .688$$

b. The critical value for r with $\alpha = .05$ and $n = 6$ is .811.

c. Since the observed correlation does not exceed this critical value, we fail to reject the null hypothesis of no correlation.

d. The slope of the regression line is given by

$$m = \frac{n\sum xy - \sum x\sum y}{n(\sum x^2) - (\sum x)^2} = \frac{6(4775) - (290)(97)}{6(14406) - (290)^2} = \frac{520}{2336} = .223.$$

To compute the y intercept, b, we need $\bar{x} = \frac{290}{6} = 48.33$ and $\bar{y} = \frac{97}{6} = 16.17$. Then $b = 16.17 - (.223)(48.33) = 5.39$. Then the regression equation is given by

$$y = .223x + 5.39.$$

e. The regression line and scatter diagram are graphed below.

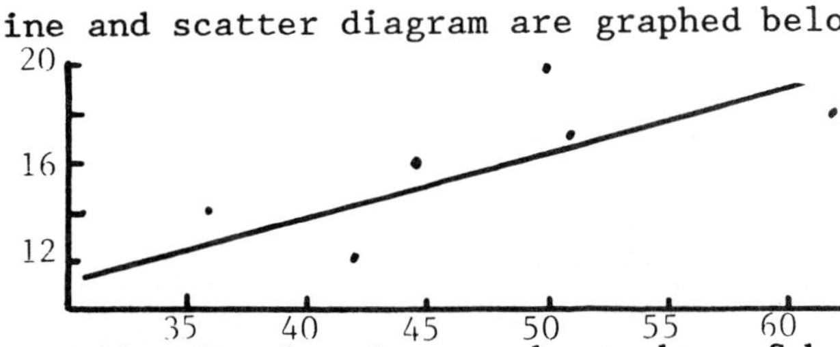

9-115 a. Setting x as the traffic level and y as the number of books sold per hour, the correlation coefficient is given by

$$r = \frac{n(\sum xy) - (\sum x)(\sum y)}{\sqrt{n(\sum x^2) - (\sum x)^2}\sqrt{n(\sum y^2) - (\sum y)^2}}$$

$$= \frac{8(12105) - (390)(211)}{\sqrt{8(6573) - (211)^2}\sqrt{8(22378) - (390)^2}} = .987.$$

b. From the table for $\alpha = .05$ and $n = 8$, the critical value for r is .707.

c. The observed correlation from the data exceeds the critical value. Therefore we reject the null hypothesis and conclude that the data supports positive linear correlation between pedestrian traffic and book sales for the store.

d. The slope of the regression line is given by

$$m = \frac{n\sum xy - \sum x\sum y}{n(\sum x^2) - (\sum x)^2} = \frac{8(12105) - (390)(211)}{8(6573) - (211)^2} = 1.8.$$

The two means are $\bar{x} = \frac{211}{8} = 26.37$ and $\bar{y} = \frac{390}{8} = 48.75$.

Then b = 48.75 - (1.8)(26.37) = 1.28.

Then the regression equation is y = 1.8x + 1.28.

e. The graph of the regression line and data is given below.

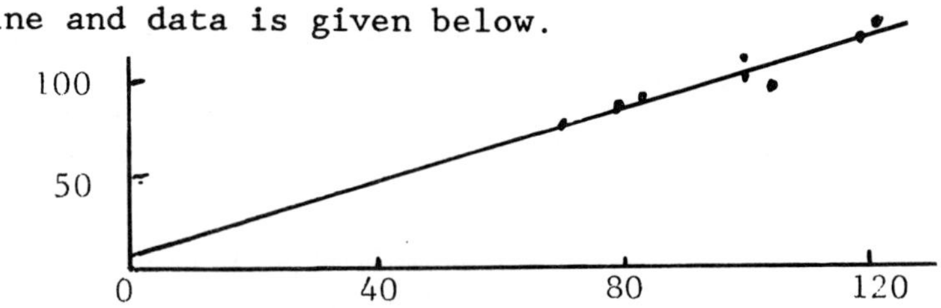

9-117 This problem is solved in a manner similar to 9-115 above.

9-119 We obtain the answers for this exercise using a computer
a. r = .884.

b. From the table the critical value is .707 for n = 8.

c. Since the observed value of r = .884 exceeds the critical value, we conclude that the data supports positive linear correlation.

d. The line is $x_2' = .177x_1 - .184$.

9-121 This exercise is solved in a similar manner as 9-119.

9-123 This exercise is solved using a computer as above.

9-125 We compute the regression equation as y' = .54 + .525x. The variations are computed using the following table. We compute $\bar{y}$ = 9/4 = 2.25.

x	y	$y-\bar{y}$	$(y-\bar{y})^2$	y'	y-y'	$(y-y')^2$
1	3	.75	.5625	1.07	1.93	3.72
2	0	-2.25	5.0625	1.59	-1.59	2.52
4	1	-1.25	1.5625	2.64	-1.64	2.69
6	5	2.75	7.5625	3.69	1.31	1.72

a.,b., and c. The unexplained variation is the sum of the last column which is 10.678. The total variation in the sum from the fourth column which is 14.75. The explained variation is the differecn.e of these two which is 4.072.

d. The coefficient of determination is the quotient of the explained variation over the total variation which is .2761.

e. The equation for the standard error of the estimate is

$$s_e = \sqrt{\frac{\Sigma(y-y')^2}{(n-2)}} = \sqrt{\frac{10.678}{2}} = 2.3106.$$

9-127 We compute the regression equation as y' = 6 - x. Also $\bar{y}$ = 11/4 = 2.75. To compute the variations we use our table of computations.

x	y	$y-\bar{y}$	$(y-\bar{y})^2$	y'	y-y'	$(y-y')^2$
4	2	-.75	.5625	2	0	0
2	4	1.25	1.5625	4	0	0
1	5	2.25	5.0625	5	0	0
6	0	-2.75	7.5625	6	0	0

a.,b.,and c. The total variation is the sum of the fourth column which is 14.75. The total unexplained variation is the sum of the final column which is 0. There is no variation which is not explained by regression. The total explained variation is therefore 14.75.

d. The coefficient of determination is the explained variation divided by the total variation. The value is 14.75/14.75 = 1.

e. Using the formula and the sum from the final column of the table, the standard error of the estimate is 0.

CHAPTER 10. CHI SQUARE AND ANALYSIS OF VARIANCE

SOLUTIONS TO SECTION 10-2 MULTINOMIAL EXPERIMENTS

10-1 a. Given 100 absences and 5 days of the week, we would expect 1/5 of the absences or 20 to occur on each day of the week. The χ^2 value based on the sample data is

$$\chi^2 = \sum\frac{(O-E)^2}{E} = \frac{(27-20)^2}{20} + \frac{(19-20)^2}{20} + \frac{(22-20)^2}{20} + \frac{(20-20)^2}{20} + \frac{(12-20)^2}{20}$$
$$= \frac{49}{20} + \frac{1}{20} + \frac{4}{20} + \frac{0}{20} + \frac{64}{20} = \frac{118}{20} = 5.9.$$

b. We have selected $\alpha = .01$. Since n = 5 (the number of days) there are 5-1 = 4 degrees of freedom. Using the .01 column of the χ^2 table we have a critical value of 13.277.

c. Because the observed χ^2 of 5.9 does not exceed the critical value of 13.277, we would fail to reject the hypothesis of equal absences throughout the week.

10-3 If the digits occur with equal frequeny, then each digit should occur 10 times. Then χ^2 is given by

$$\chi^2 = \sum\frac{(O-E)^2}{E} = \frac{(8-10)^2}{10} + \frac{(8-10)^2}{10} + \frac{(12-10)^2}{10} + \frac{(11-10)^2}{10} +$$
$$\frac{(10-10)^2}{10} + \frac{(8-10)^2}{10} + \frac{(9-10)^2}{10} + \frac{(8-10)^2}{10} + \frac{(12-10)^2}{10} + \frac{(14-10)^2}{10}$$
$$= 4.2.$$

There are 9 degrees of freedom. The critical value of χ^2 at $\alpha = .05$ is 16.919. Since the sample value of $\chi^2 = 4.2$, does not exceed the critical value, we fail to reject the null hypothesis of equality of frequency of the digits.

10-5 If all proportions are equal, then each of the five groups should have 30 people. Then

$$\chi^2 = \sum\frac{(O-E)^2}{E} = \frac{(36-30)^2}{30} + \frac{(32-30)^2}{30} + \frac{(35-30)^2}{30} + \frac{(31-30)^2}{30} +$$
$$\frac{(16-30)^2}{30} = 8.73.$$

There are 4 degrees of freedom because of the 5 classes. The critical value of χ^2 at $\alpha = .01$ is 13.277. The value of χ^2 from the data is 8.73 which does not exceed the critical value. Therefore, we fail to reject the null hypothesis of equality of proportions.

10-7 If the 8 brands were selected with equal frequency, then each brand would have been selected as best by 15 of the 120 subjects. Consequently the expected frequencies are 15 for each brand. We form χ^2 by

$$\chi^2 = \sum\frac{(O-E)^2}{E} = \frac{(10-15)^2}{15} + \frac{(14-15)^2}{15} + \frac{(16-15)^2}{15} + \frac{(21-15)^2}{15} +$$
$$\frac{(17-15)^2}{15} + \frac{(14-15)^2}{15} + \frac{(15-15)^2}{15} + \frac{(13-15)^2}{15} = \frac{72}{15} = 4.8.$$

There are 8 cells, so there are 8-1 = 7 df. Using $\alpha = .05$, the critical χ^2 value is 14.067. Since the observed χ^2 value does not exceed this critical value, we fail to reject the hypothesis of equal effectiveness of the 8 brands of aspirin.

10-9 There are a total of 103 shoppers in the sample. If the percentages are correct, then the expected numbers by day are: Sun 7.21, Mon 5.15, Tue 9.27, Wed 11.33, Thu 19.57, Fri 24.72, Sat 25.75. Computing χ^2 we would have

$$\chi^2 = \sum \frac{(O-E)^2}{E} = \frac{(9-7.21)^2}{7.21} + \frac{(6-5.15)^2}{5.15} + \frac{(10-9.27)^2}{9.27} + \frac{(8-11.33)^2}{11.33} + \frac{(19-19.57)^2}{19.57} + \frac{(23-24.72)^2}{24.72} + \frac{(28-25.75)^2}{25.75}$$

Adding these terms up we arrive at a χ^2 observed of 1.9. There are 7 cells and so there are 6 df. We have $\alpha = .05$. This will give a critical χ^2 value of 12.592. Since the observed χ^2 value does not exceed this critical value, we fail to reject the hypothesis that the percentages are correct.

10-11 This problem is solved in a manner similar to 11-7. If each month is equally likely, then the 120 selections would be divided equally among the 12 months giving 10 each. We form the χ^2 value and compare it against the critical χ^2 value of 19.675. We would reject or accept the null hypothesis based upon comparison with this critical value.

10-13 There are a total of 270 people surveyed. If the percentages are correct, then the expected number of samples in the cells would be:
light red/brown 59.4, white 59.4, light blue 54, dk. blue 48.6, red 48.6. Then χ^2 is given by

$$\chi^2 = \sum \frac{(O-E)^2}{E} = \frac{(60-59.4)^2}{59.4} + \frac{(61-59.4)^2}{59.4} + \frac{(43-54)^2}{54} + \frac{(41-48.6)^2}{48.6} + \frac{(65-48.6)^2}{48.6} = 9.02$$

There are 4 degrees of freedom. The critical value of χ^2 with $\alpha = .05$ is 9.488. The observed value of χ^2 does not exceed this so we fail to reject the null hypothesis that the percentages are correct.

10-15 There are a total of 150 respondents. If the given percentages are correct, then the expected number of respondents should be: A 33, B 27, C 18, D 18, E 15, F 13.5, G 13.5, H 12. Then we compute χ^2 as follows.

$$\chi^2 = \sum \frac{(O-E)^2}{E} = \frac{(29-33)^2}{33} + \frac{(30-27)^2}{27} + \frac{(20-18)^2}{18} + \frac{(16-18)^2}{18} + \frac{(9-15)^2}{15} + \frac{(17-13.5)^2}{13.5} + \frac{(10-13.5)^2}{13.5} + \frac{(19-12)^2}{12}.$$

Adding these , we get an observed χ^2 of 9.56. There are 7 degrees of freedom. The critical value of χ^2 is 14.067. Since the observed value of χ^2 is not in the critical region, we fail to reject the null hypothesis that the results from the second consultant are the same as the first.

10-17 Using the given percentages, the expected number of people registering will be: Mon 740, Tue 832.5, Wed 277.5. The computed value of χ^2 is

$$\chi^2 = \sum \frac{(O-E)^2}{E} = \frac{(1073-740)^2}{740} + \frac{(555-832.5)^2}{832.5} + \frac{(222-277.5)^2}{277.5} = 253.5.$$

The critical value based upon $\alpha = .01$ and 2 degrees of freedom is 9.21. Since the observed χ^2 exceeds the critical value, we reject the hypothesis that the data matched the deans model.

10-19 If each of the 36 possible outcomes is to have an expected frequency of 5, there must be 36 x 5 = 180 throws. We know from the χ^2 table that for 36 outcomes or 35 degrees of freedom at $\alpha = .01$, the critical value is between 50.892, which is the critical value for 30 df, and 63.691, which is the critical value for 40 df. Since the observed χ^2 exceeds any number in this range, we would reject the hypothesis of equal frequencies.

SOLUTIONS TO 10-2 B EXERCISES

10-23 There are a number of ways to combine these columns having expected frequencies under 5. Most commonly one simply combines a column with an adjacent one. We combine as in the following table:

Time Slot	1+2	3	4	5+6	7+8	9+10
Observed	10	8	9	8	3	15
Expected	9	8	7	10	7	12

To obtain the 1+2 time slot, we have the observed number given by 2+8, where we have added the observed frequency from column 1 to that of column 2. Similarly we combined the 4 and 5 to get an expected frequency of 9. Other combined cells were generated in a similar manner. The observed χ^2 is

$$\chi^2 = \frac{(10-9)^2}{9} + \frac{(8-8)^2}{8} + \frac{(9-7)^2}{7} + \frac{(8-10)^2}{10} + \frac{(3-7)^2}{7} + \frac{(15-12)^2}{12} =$$

$$= .11 + .57 + .4 + 2.29 + .75 = 4.12.$$

There are 5 df because the combined table has only 6 classes. The critical value of χ^2 using $\alpha = .05$ is 11.071. Since the observed χ^2 does not exceed this, we fail to reject the model of expected numbers of observations as stated.

SOLUTIONS TO SECTION 10-3 CONTINGENCY TABLES

10-25 a. We establish row and column totals by:

	In Favor	Opposed	
Democrat			50
Republican			40
	50	40	

There is a grand total of 90. Using the formula for expected frequencies yields

	In Favor	Opposed
Democrat	$\frac{50 \times 50}{90} = 27.778$	$\frac{50 \times 40}{90} = 22.222$
Republican	$\frac{40 \times 50}{90} = 22.222$	$\frac{40 \times 40}{90} = 17.778$

where each expected frequency is obtained by

$$\frac{\text{(Row Total) x (Column Total)}}{\text{(Grand Total)}}$$

b. The observed χ^2 value is

$$\chi^2 = \frac{(27.778-20)^2}{27.778} + \frac{(22.222-30)^2}{22.222} + \frac{(22.222-30)^2}{22.222} + \frac{(17.778-10)^2}{17.778}$$

$$= 11.02.$$

c. The critical value based upon α = .05 and (rows - 1) x (columns - 1) = 1 x 1 = 1 df is 3.841.

d. Since the observed χ^2 of 11.02 exceeds the critical value, we would reject the hypothesis of independence of party affiliation and opinion on this gun control bill.

10-27 The row sums for group A and group B respectively are 24 and 562. The column sums for pass and fail respectively are 427 and 159. The grand total is 588. We construct the table of expected frequencies as follows.

	Pass	Fail
Group A	$\frac{427 \times 24}{588} = 17.4$	$\frac{159 \times 24}{588} = 6.6$
Group B	$\frac{427 \times 562}{588} = 408$	$\frac{159 \times 562}{588} = 154$

We compute χ^2 by

$$\chi^2 = \frac{(17.4-10)^2}{17.4} + \frac{(6.6-14)^2}{6.6} + \frac{(408-417)^2}{408} + \frac{(154-145)^2}{154}$$

$$=12.3.$$

We have 1 degree of freedom. Using α = .05, the critical value of χ^2 is 3.841. Since our observed χ^2 exceeds this critical value, we would reject the hypothesis of independence of these two variables.

10-29 We make a table below containing both row and column totals as well as expected frequencies as are given in the problem.

	Control	Experimental	Totals
Medicare	$\frac{81 \times 65}{135} = 39$	$\frac{81 \times 70}{135} = 42$	81
Medicade	$\frac{33 \times 65}{135} = 15.9$	$\frac{33 \times 70}{135} = 17.1$	33
Welfare	$\frac{21 \times 65}{135}$ 10.1	$\frac{21 \times 70}{135} = 10.9$	21
Totals	65	70	135

The grand total is 135. The observed χ^2 value is

$$\chi^2 = \frac{(38-39)^2}{39} + \frac{(43-42)^2}{42} + \frac{(16-15.9)^2}{15.9} + \frac{(17-17.1)^2}{17.1} + \frac{(11-10.1)^2}{10.1} + \frac{(10-10.1)^2}{10.1} = .216.$$

Using α = .05 and degrees of freedom equal to (r-1)x(c-1) = 2 we have a critical value of 5.99. Since the observed χ^2 does not exceed this, we fail to reject the hypothesis of independence between payerand group.

10-31 This problem is solved with a contingency table in a manner similar to 10-27.

10-33 The contingency table of expected numbers is contained below.

Day	Mon	Tue	Wed	Thurs	Fri	Totals
Acceptable	86.5	95.6	91.1	91.1	85.7	450
Defective	8.5	9.4	8.9	8.9	8.3	44
Totals	95	105	100	100	94	494

The total value of χ^2 is

$$\chi^2 = \frac{6.5^2}{86.5} + \frac{4.4^2}{95.6} + \frac{3.9^2}{91.1} + \frac{1.9^2}{91.1} + \frac{3.7^2}{85.7} + \frac{6.5^2}{8.5} + \frac{4.4^2}{9.4} + \frac{3.9^2}{8.9}$$

$$+ \frac{1.9^2}{8.9} + \frac{3.7^2}{8.3} = 11.85.$$

There are 5 columns and 2 rows and so the are (5-1)(2-1) = 4 degrees of freedom. The critical value using $\alpha = .05$ is 9.499. Therefore, we reject the null hypothesis of independence of numbers of defectives and day of the week. The data supports a conclusion of dependence.

10-35 We generate a contingency table for the test of independence. The expected values are given in the following table.

	Natural	Motor Veh.	Suicide/Homocide	Totals
Death Cert.	5.7	8.7	5.6	20
Med Panel	33.3	51.4	33.3	118
Totals	39	60	39	138

Then χ^2 is given by

$$\chi^2 = \frac{3.3^2}{5.7} + \frac{1.7^2}{8.7} + \frac{1.6^2}{5.6} + \frac{3.3^2}{33.3} + \frac{1.6^2}{51.4} + \frac{1.7^2}{33.3} = 3.16.$$

There are 2 degrees of freedom for this test. Using $\alpha = .05$, the critical value is 5.991. Since the observed χ^2 does not exceed the critical value, we accept the null hypothesis of independence of the two variables.

10-37 This exercise is solved in the same manner as 10-35.

10-39 This is a larger table having three rows and five columns. There is a grand total of 754 students. We form a table of row and column sums plus expected values in the usual fashion.

Course	Grade					
	A	B	C	D	F	Total
Math	16.8	31.6	30.8	7.6	13.2	100
Art	30.1	56.5	55.1	13.5	35.8	179
English	80.1	149.9	146.1	35.9	51.	475
Totals	127	238	232	57	100	

Each entry above is generated using the formula row sum times column sum divided by the grand total. The resultant value of χ^2 using all 15 terms is 17.3. In this case there are 8 df. Using a significance level of $\alpha = .10$, the critical value of χ^2 is 13.362. Since our observed χ^2 exceeds this, we would reject a hypothesis of independence of grades and subject.

SOLUTIONS TO 10-3 B EXERCISES

10-41 One reasonable way of combining rows would be on a geographic basis. The resultant table would be as shown below.

	Favor	Opposed	No Opinion
East	11	16	9
Central	11	7	9
West	14	34	7

We obtained the 11 in the East in favor for example by combining the 4 of the Northeast with the 7 of the Southeast. The expected values along with row and column sums are given in the table below.

	Favor	Opposed	No Opinion	Totals
East	11	17.4	7.66	36
Central	8.2	13.0	5.7	37
West	16.8	26.6	11.7	55
Total	36	57	25	

Expected frequencies are

	Favor	Opposed	No Opinion
East	$\frac{36x36}{118} = 11$	$\frac{57x36}{118} = 17.4$	$\frac{25x36}{118} = 7.6$
Central	$\frac{27x36}{118} = 8.2$	$\frac{27x57}{118} = 13.0$	$\frac{27x25}{118} = 5.7$
West	$\frac{55x36}{118} = 16.8$	$\frac{55x27}{118} = 26.6$	$\frac{55x25}{118} = 11.7$

Then $\chi^2 = \frac{(11-11)^2}{11} + \frac{(16-17.4)^2}{17.4} + \frac{(9-7.6)^2}{7.6} + \frac{(11-8.2)^2}{8.2} + \frac{(7-13)^2}{13} +$

$+ \frac{(9-5.7)^2}{5.7} + \frac{(14-16.8)^2}{16.8} + \frac{(34-26.6)^2}{26.6} + \frac{(7-11.7)^2}{11.7} = 10.36.$

Using $\alpha = .05$ with $(r-1)(c-1) = (3-1)(3-1) = 4$ degrees of freedom, there is a critical χ^2 value of 9.488. Since our observed χ^2 value exceeds this, we reject the hypothesis of independence of geographic region and opinion.

SOLUTIONS TO SECTION 10-4 ANALYSIS OF VARIANCE

10-45 The variance within is given by

$$\frac{s^2_{res} + s^2_{com} + s^2_{pt}}{3} = \frac{.3 + .25 + .16}{3} = .24.$$

For the variance between we use the means. For the overall mean,

$$\bar{x} = \frac{2.6 + 2.55 + 2.3}{3} = 2.48.$$

Then

$$s^2_{\bar{x}} = \frac{(\bar{x}_r - 2.48)^2 + (\bar{x}_c - 2.48)^2 + (\bar{x}_{pt} - 2.48)^2}{2}$$

$= .026.$

Then F is given by $F = (15)(.026)/.24 = 1.64$, because there are 15 samples in each of the groups. The critical F, using $\alpha = .05$, a numerator degrees of freedom of 3-1=2, and a denominator df of 3(15-1) = 42, F is at least 3.1. Since the observed F does not exceed this, we fail to reject the hypothesis of equality of the means.

10-47 We find the variance within as

$$s_w^2 = \frac{s_A^2 + s_B^2 + s_C^2 + s_D^2 + s_E^2}{5} = 151.$$

For the variance between we compute the overall mean.

$$\bar{x} = \frac{103 + 97 + 102 + 100 + 110}{5} = 102.4.$$

The variance between is then

$$s_b = \frac{(103-102.4)^2 + (97-102.4)^2 + (102-102.4)^2 + (100-102.4)^2 + (110-102.4)^2}{4}$$

$= 23.3$

Then F is given by $F = \frac{(10)(23.3)}{151} = 1.53$. There are 5-1 = 4 df in the numerator and k(n-1) = 5(10-1) = 45 df in the denominator. Using α = .05, the critical value of F is more than 2.60. Since the observed F does not exceed any such number, we fail to reject a hypothesis that the means are equal.

10-49 We need the means and variances from the raw data for each of the car models. These are computed using the formula of chapter 2.
Then

$$\bar{x}_A = 18, \ \bar{x}_B = 19, \ \bar{x}_C = 19, \ \bar{x}_D = 21, \ \bar{x}_E = 17,$$

and

$$s_A^2 = 7.3, \ s_B^2 = 11.3, \ s_C^2 = 1.33, \ s_D^2 = 1.33, \ s_E^2 = 4.67.$$

Then the variance within is the average of these 5 variances.

$$s_w^2 = \frac{7.3 + 11.3 + 1.33 + 1.33 + 4.67}{5} = 5.2.$$

We compute the variance between by first finding the overall mean of the above 5 means. That mean is 18.8. The variance between is then

$$s_b^2 = \frac{(18-18.8)^2 + (19-18.8)^2 + (19-18.8)^2 + (21-18.8)^2 + (17-18.8)^2}{4}$$

$= 2.2.$

The observed F uses n = 4 cars per model.

$$F = \frac{ns^2_b}{s^2_w} = \frac{(4)(2.2)}{5.2} = 1.7.$$

We compute the critical F using α = .05, and k = 5 models. Then the numerator df is k-1 = 4, and the denominator df is 5(4-1) = 15. The critical value of F is 3.0556. The observed F does not exceed this; therefore we fail to reject the hypothesis of equality of the means.

10-51 This exercise is solved in the same manner as 10-49.

10-53 We have unequal numbers of observations in the samples. As a result we follow the format of the ANOVA Table. The claim is that all three samples come from populations having equal means.

Number of samples is k = 3

Number of values is $n_1 + n_2 + n_3 = 5 + 7 + 6 = 18$

The column sums are $c_1 = 437$, $c_2 = 484$, and $c_3 = 357$

$\sum x = \sum c_i = 437 + 484 + 357 = 1278$

$(\sum x)^2 = (1278)^2 = 1633284$

$\sum x^2 = 93926$

$SS(total) = \sum x^2 - (\sum x)^2/N = 93926 - (1633284)/18 = 3188$

$$SS(treatment) = \frac{437^2}{5} + \frac{484^2}{7} + \frac{357^2}{6} - \frac{1278^2}{18} = 2162$$

$SS(error) = SS(total) - SS(treatment) = 3188 - 2162 = 1026$
$df(treatment) = k - 1 = 3 - 1 = 2$
$df(error) = N - k = 18 - 3 = 15$

$$MS(treatment) = \frac{SS(treatment)}{df(treatment)} = \frac{2162}{2} = 1081$$

$$MS(error) = \frac{SS(error)}{df(error)} = \frac{1062}{15} = 70.8$$

$$F = \frac{MS(treatment)}{MS(error)} = \frac{1081}{70.8} = 15.27.$$

The significance level indicates an $\alpha = .05$. There are 2 degrees of freedom in the numerator and 15 in the denominator. The critical value for F is 3.68. Since the observed F exceeds this, we reject the hypothesis of equality of the means.

10-55 We use the method as described in the book. The null hypothesis is that the three means are equal. The alternate hypothesis is that they are not all equal.
Number of samples = 3
Total number of values = 4 + 6 + 8 = 18
c_1 = sum of column A = 72
c_2 = sum of column B = 96
c_3 = sum of column C = 162

$\sum x$ = overall sum = 72 + 96 + 162 = 330

$(\sum x)^2 = 330^2 = 108900$

$\sum(x^2) = 6160$

$SS(total) = 6160 - (108900)/18 = 110$

$$SS(treatment) = \sum \frac{c^2_i}{n_i} - \frac{(\sum c_i)^2}{N} = 6112.5 - 6050 = 62.5$$

$SS(error) = SS(total) - SS(treatment) = 110 - 62.5 = 47.5.$

$df(treatment) = 3 - 1 = 2$
$df(error) = N - k = 18 - 3 = 15$

$$MS(treatment) = \frac{SS(treatment)}{df(treatment)} = \frac{62.5}{2} = 31.25$$

$$MS(error) = \frac{SS(error)}{df(error)} = \frac{47.5}{15} = 3.17$$

$$F = \frac{MS(treatment)}{MS(error)} = \frac{31.25}{3.17} = 9.86.$$

We use a significance level of .05. The numerator degrees of freedom is 2 and the denominator df is 15. The critical value of F is 3.68. Since the computed F exceeds this level, we reject the hypothesis that the means are equal.

10-57 The null hypothesis is that the three recovery times are from populations with equal means. The alternate is that they are from populations whose means are not all equal.
Number of samples = 3
Total number of values = 6 + 5 + 8 = 19
$c_1 = 44$
$c_2 = 39$
$c_3 = 84$

$\sum x$ = overall sum = 44 + 39 + 84 = 167

$(\sum x)^2 = 27889$

$\sum(x^2) = 1597$

SS(total) = 1597 - (27889)/19 = 129.2

$$SS(treatment) = \frac{44^2}{6} + \frac{39^2}{5} + \frac{84^2}{8} - \frac{167^2}{19} = 41$$

SS(error) = SS(total) - SS(treatment) = 129.2 - 41 = 88.2
df(treatment) = 2
df(error) = 19 - 3 = 16

$$MS(treatment) = \frac{SS(treatment)}{df(treatment)} = \frac{41}{2} = 20.5$$

$$MS(error) = \frac{SS(error)}{df(error)} = \frac{88.2}{16} = 5.51$$

$$F = \frac{MS(treatment)}{MS(error)} = \frac{20.5}{5.51} = 3.72.$$

There are 2 df for the numerator and 16 for the denominator. Using $\alpha = .01$, we have a critical value of 6.226. Since the computed value of F does not exceed this, we fail to reject the hypothesis of equal means.

10-59 This problem is solved in a manner similar to 10-57 above.

10-61 We list the data in columns below

	Zone 1	Zone 4	Zone 7
	147	160	215
	160	140	127
	128	173	98
	162	113	147
	135	85	184
	132	120	109
	181	285	169
	138	117	110
	145	133	68
	165	119	160
Totals	1493	1445	1387

The null hypothesis is that the three zones have the same mean selling price. The alternate hypothesis is that there are some differences. We follow the outline to determine F.

k = 3 samples
n_i = 10 for all three samples
The column sums c_i are given above
N = 30
$\sum x = 1493 + 1445 + 1387 = 4325$
$\sum(x^2) = 672017$
$SS(total) = 672017 - (4325)^2/30 = 48496$

$$SS(treatment) = \frac{1493^2}{10} + \frac{1445^2}{10} + \frac{1387^2}{10} - \frac{4325^2}{30} = 563.$$

$SS(error) = 48496 - 563 = 47933.$
$df(treatment) = 3 - 1 = 2$
$df(error) = 30 - 3 = 27$

$$MS(treatment) = \frac{563}{2} = 281.5$$

$$MS(error) = \frac{47933}{27} = 1775$$

$$F = \frac{281.5}{1775} = .16.$$

The critical value for F with 2 df in the numerator and 27 in the denominator and $\alpha = .05$ is 3.3541. Since the observed F does not exceed this, we fail to reject the null hypothesis of equality of the variances.

10-63 This exercise is solved in the same manner as 10-61.

10-65 This exercise is solved in the same manner as 10-45.

10-67 This exercise is solved in the same manner as 10-47.

SOLUTIONS TO 10-4 B EXERCISES

10-69 Since each sample size in the three precincts exceeds 30 we use the normal distribution

$$z = \frac{\bar{x}_A - \bar{x}_B}{\sqrt{\frac{s_A^2}{n_A} + \frac{s_B^2}{n_B}}}.$$

a. Our null hypothesis is $H_0: \mu_1 = \mu_2$. The alternate hypothesis is that they are not equal. Computing z we have

$$z = \frac{170 - 202}{\sqrt{\frac{324}{50} + \frac{400}{50}}} = \frac{-32}{3.81} = -8.4.$$

The critical values for a two tailed test with $\alpha = .05$ are 1.96 and -1.96. Since the computed z is in the critical region, we reject the null hypothesis that the means are equal.

b. Our null hypothesis is $H_0: \mu_2 = \mu_3$. The alternate hypothesis is that they are not equal. Computing z we get

$$z = \frac{202 - 165}{\sqrt{\frac{400}{50} + \frac{529}{50}}} = \frac{37}{4.31} = 8.58.$$

The critical values are the same as in a above. Since the observed z is in the critical region, we reject the hypothesis that the two means are equal.

c. Our null hypothesis is H_0: $\mu_1 = \mu_3$. The alternate is that they are not equal. Computing z we get

$$z = \frac{170 - 165}{\sqrt{\frac{324}{50} + \frac{529}{50}}} = \frac{5}{4.13} = 1.21.$$

Since this value of z is not in the critical region, we fail to reject the hypothesis of equality of the two means.

d. Using ANOVA with equal sample sizes we have

$$\text{Variance within} = \frac{(324 + 400 + 529)}{3} = 418.$$

$$\text{The overall mean} = \frac{170 + 202 + 165}{3} = 179.$$

The variance of the means is

$$\frac{(170 - 179)^2 + (202 - 179)^2 + (165 - 179)^2}{2} = 403.$$

The variance between is $ns^2 = (50)(403) = 20150$.
Then F is the quotient of these two variances = (20150)/418 = 48.2.
The numerator df = k - 1 = 2.
The denominator df = k(N-1) = 3(49) = 147.
Using $\alpha = .05$, the critical value of F is less than 3.92. Since the observed F exceeds this, we reject the hypothesis of equality.

e. The ANOVA causes us to believe that the means are not equal with a 5% risk of type one error. The three tests suggest which means are not equal. However the risk of at least one error in the three tests each having a 5% chance of type one error is greater.

SOLUTIONS TO REVIEW EXERCISES

10-73 Using a contingency table, we have the following table of expected responses

	In Favor	Against	No Opinion
Northeast	8.5	21.5	5.
South	13.5	34.5	8.

In each case we computed the value by row sum x column sum divided by the total number of 91. We compute χ^2 as

$$\chi^2 = \frac{(10-8.50)^2}{8.5} + \frac{(12-13.5)^2}{13.5} + \frac{(20-21.5)^2}{21.5} + \frac{(36-34.5)^2}{34.5} + \frac{(5-5)^2}{5}$$

$$+ \frac{(8-8)^2}{8} = .6.$$

The critical χ^2 value of 2 degrees of freedom and $\alpha = .05$ is 5.991. Since the observed χ^2 does not exceed this critical level, we fail to reject the hypothesis of independence of opinion and geographic region.

10-75 We use ANOVA on this problem. The sample sizes for the three groups are equal.

Therefore, we compute
Variance within = $((1.4)^2 + (1.1)^2 + (1.5)^2)/3 = 1.81$.
The overall mean is $(5.2 + 4.1 + 5.5)/3 = 4.9$.
The variance of the means is .545.
The variance between is $(36)(.545) = 19.62$.
F is the quotient of these two variances = $19.62/1.81 = 10.84$.
Numerator df = $k - 1 = 2$.
Denominator df = $k(n-1)) = 3(35) = 105$.
With an $\alpha = .01$ the critical value of F is less than 3.077. Since the computed F exceeds this, we reject the hypothesis of equality among the judges.

10-77 There is a total of 560 customers during the 7 days. This amounts to an expected number of 80 per day. The null hypothesis is that all days had equal numbers of customers. The test statistic is χ^2.

$$\chi^2 = \sum \frac{(o - e)^2}{e} = \frac{(97-80)^2}{80} + \frac{(72-80)^2}{80} + \frac{(55-80)^2}{80} + \frac{(68-80)^2}{80} +$$

$$\frac{(70-80)^2}{80} + \frac{(88-80)^2}{80} + \frac{(110-80)^2}{80} = 27.33.$$

There are 7 cells so there are 7-1 = 6 degrees of freedom. The critical χ^2 value using $\alpha = .05$ is 12.592. Since the observed χ^2 value exceeds this critical level, we reject the hypothesis that the expected number of customers is the same for each day of the week.

10-79 To test independence, we will use a contingency table. We compute expected frequencies in the following table using row sums of 59 and 47 for the top and bottom rows and column sums for A-F respectively of 19, 31, 28, 13, 15.

	A	B	C	D	F
Math	$\frac{59x19}{106}$	$\frac{59x31}{106}$	$\frac{59x28}{106}$	$\frac{59x13}{106}$	$\frac{59x15}{106}$
English	$\frac{47x31}{106}$	$\frac{47x31}{106}$	$\frac{47x28}{106}$	$\frac{47x13}{106}$	$\frac{47x15}{106}$

The expected frequencies are then

	A	B	C	D	F
Math	10.6	17.3	15.6	7.2	8.3
English	8.4	13.7	12.4	5.8	6.7

Then χ^2 is given by

$$\chi^2 = \frac{(10.6-13)^2}{10.6} + \frac{(8.4-6)^2}{8.4} + \frac{(17.3-16)^2}{17.3} + \frac{(13.7-15)^2}{13.7} + \frac{(15.6-13)^2}{15.6} +$$

$$\frac{(12.4-15)^2}{12.4} + \frac{(7.2-8)^2}{7.2} + \frac{(5.8-5)^2}{5.8} + \frac{(8.3-9)^2}{8.3} + \frac{(6.7-6)^2}{6.7}$$

$$= 2.76.$$

The critical χ^2 value for $(r-1)(c-1) = (1)(4) = 4$ degrees of freedom is 9.488 using $\alpha = .05$. Since the observed χ^2 is not in the critical region, we fail to reject the null hypothesis of independence of distribution of grades between the two subjects.

10-81 We will use an analysis of variance to test the hypothesis of equality of the three means. The alternative hypothesis is that there is some inequality. We follow the format of the book.

k = 4 samples
The respective sample sizes are 5,5,5, and 10.
$c_1 = 53$, $c_2 = 64$, $c_3 = 62$, and $c_4 = 227$.
N = 25.
$\sum x = 406$.
$\sum(x^2) = 8454$.
SS(total) = 8454 - $(406)^2/25$ = 1860.6

$$SS(\text{treatment}) = \frac{53^2}{5} + \frac{64^2}{5} + \frac{62^2}{5} + \frac{227^2}{10} - \frac{406^2}{25} = 709.26.$$

SS(error) = 1860.6 - 709.3 = 1151.3
df(treatment) = 4 - 1 = 3
df(error) = 25 - 4 = 21

$$MS(\text{treatment}) = \frac{709.26}{3} = 236.4$$

$$MS(\text{error}) = \frac{1151.3}{21} = 54.8$$

$$F = \frac{MS(\text{treatment})}{MS(\text{error})} = 4.31.$$

From the table the critical value for F with α = .01 is 4.874. Since the observed value of F does not exceed the critical value, we fail to reject the null hyopthesis of equality of the means.

10-83 The null hypothesis is that the 4 characteristics are equally likely. The alternate hypothesis is that they are not. If they are equally likely, then we would expect 25 subjects to have each characteristic. Using the χ^2 statistic, we have

$$\chi^2 = \sum \frac{(o - e)^2}{e} = \frac{(20-25)^2}{25} + \frac{(30-25)^2}{25} + \frac{(15-25)^2}{25} + \frac{(35-25)^2}{25}$$
$$= 10.$$

With α = .01 and 3 degrees of freedom, the critical value for χ^2 is 11.345. Since the observed χ^2 value does not exceed this, we fail to reject the null hypothesis that the characteristics are equally likely.

CHAPTER 11. NON-PARAMETRIC STATISTICS

SOLUTIONS TO SECTION 11.2 - SIGN TEST

11-1 We form the table for the sign test by using a + sign if the sitting force exceeds the supine and a - if it is less

sitting	4.66	5.70	5.37	3.34	3.77	7.43	4.15	6.21	5.90	5.77
supine	4.63	6.34	5.72	3.23	3.60	6.96	3.66	5.81	5.61	5.33
sign	+	-	-	+	+	+	+	+	+	+

There are 8 + signs in 10 samples and 2 - signs. Therefore x = 2 (the smaller of 2 and 8). From the table of critical values of the sign test the critical region is 1 or fewer - signs. Therefore, we fail to reject the hypothesis of no difference in the two positions.

11-3 We form the sign test by a plus sign if $y > x$, 0 if $y = x$, and minus if $y < x$.

x	1	2	2	3	5	6	8	4	6	7	2	5	3	2
y	2	0	1	4	4	4	9	3	5	7	2	6	2	1
Sign	+	-	-	+	-	-	+	-	-	0	0	+	-	-

There are 4 plus signs and 8 minus signs out of the twelve signs counted. Since we are testing that they come from the same population, we use a two-tailed test with n = 12. The critical number is 2. Since the 4 plus signs exceed this, we fail to reject the null hypothesis. The evidence supports their being from the same population.

11-5 Our claim is that there is no change in blood pressure due to the pill. That is

H_0 : There is no change in blood pressure due to the pill

H_1 : There is a decrease in blood pressure due to the pill

Before Pill Blood Pressure	After Pill Blood Pressure	Sign of Change from Before to After
120	118	-
136	122	-
160	143	-
98	105	+
115	98	-
110	98	-
180	180	0
190	175	-
138	105	-
128	112	-

There are 8 decreases in blood pressure, 1 increase, and 1 no change. In the sign test then, n = 9 because the one no change is discarded. The critical value using $\alpha = .05$ is x = 1 from the table. Since the smaller of 1 and 8, which is the 1, is less than or equal to the critical value, we reject the hypothesis H_0 in favor of H_1. The data supports a conclusion that the pill does decrease blood pressure.

11-7 A test of no effect against a claim of improved intelligence would be a test of the form

H_0 : Eating fish does not change intelligence

H_1 : Eating fish increases intelligence

Our table of signs is

IQ before diet	98	110	105	121	100	88	112	92	99	109	103	104
IQ after diet	98	112	106	118	102	97	115	90	99	110	105	109
Sign	0	+	+	-	+	+	+	-	0	+	+	+

There were 8 increases, 2 decreases and 2 no changes. Then n = 10 and x = 2. From the table for a one-tailed test with α = .05, the critical value from Table A-7 is 1. Since we observed x = 2, which exceeded this, we failed to reject H_0. Evidence supports that eating fish does not change intelligence.

11-9 We are testing a claim of no difference. Then

H_0 : There is no difference in ages

H_1 : There is some difference in ages

Our table is

Husband	28.1	33.0	29.8	53.1	56.7	41.6	50.6	21.4	62.0	19.7
Wife	28.4	27.6	32.7	52.0	58.1	41.2	50.7	20.6	61.1	18.1
Sign	-	+	-	+	-	+	-	+	+	+

There are 6 plus signs and 4 minus signs making n = 10 and x = 4. The critical number for n = 10 and α = .01 from the table is 0. Since the smaller of 6 and 4 is x = 4, which exceeds 0, we fail to reject H_0. Evidence supports no difference in ages.

11-11 The null hypothesis is that there is no difference in pain level before and after the hypnosis. We form the table for the sign test as follows putting a + sign if After exceeds Before

Before	-5.5	-5.0	-6.6	-9.7	-4.0	-7.0	-7.0	-8.4
After	-1.4	-0.5	0.7	1.0	2.0	0.0	-0.6	-1.8
Sign	+	+	+	+	+	+	+	+

There are 8 + signs and 0 - signs. At a .05 level of significance the critical value is 0. Therefore we reject the null hypothesis of no difference in pain level. The evidence supports the hypothesis of a difference in pain level before and after hypnosis.

11-13 Since the claim is that more than half understood, we would have

H_0 : Half or fewer understand

H_1 : More than half understand

We denote those who understand with a + sign and those who do not with a minus sign. In our case n = 20, which is fewer than 25, and so we use the sign test. This is a one-tailed test with α = .05, and hence from the table the critical value is 5. Since x = 5 did respond negatively, we would find that the data supports H_1 that more than half understand.

11-15 Using a + sign to denote Republican preference, there are 7 plus signs and 13 minus signs. Then we are testing

H_0 : The parties are preferred equally

H_1 : The parties are not preferred equally

For this two-tailed test with $\alpha = .05$, the critical number from the table is 5. Since x = 7 is not in the critical region of 5 or fewer, we fail to reject the null hypothesis that the two parties respond equally.

11-17 Since the sample size is n = 75, which is more than 25, we will use a z test. Our test is

H_0 : the median is at least 40

H_1 : the median is less than 40

This is a one-tailed test with x = 32. Then we compute

$$z = \frac{(x + 0.5) - n/2}{\sqrt{n}/2} = \frac{32.5 - 37.5}{\sqrt{75}/2} = \frac{-5}{4.33} = -1.15.$$

For a left-tailed test with $\alpha = .05$, the critical value is -1.64. Since -1.15 is not less than that value, we fail to reject H_0. The data does not support a conclusion of a median less than 40.

11-19 Since we are using the sign test but n is greater than 25, we will use the normal approximation

$$z = \frac{(x + 0.5) - n/2}{\sqrt{n}/2}.$$

Our claim is that production increased and thus we have

H_0 : Production was no better

H_1 : Production increased

We would conclude H_1 if a significantly large number of locations showed increases in production. We compute the observed value of z by using the lower x = 18.

$$z = \frac{(18 + 0.5) - (50/2)}{\sqrt{50}/2} = \frac{18.5 - 25}{\sqrt{50}/2} = -1.84.$$

This is a one-tailed test on a normal variable, and hence the critical point is -1.64. Since the observed z is less than -1.64, we would reject H_0 and conclude that the data favors H_1 that production increased.

11-21 Since there were 28 persons with different reaction times, we use n = 28. Then we can use the normal distribution in the test.

H_0 : There is no change in reaction times

H_1 : There is a change in reaction times

We compute using x = 6 our value of

$$z = \frac{(x + 0.5) - n/2}{\sqrt{n}/2} = \frac{6.5 - 14}{\sqrt{28}/2} = \frac{-7.5}{2.65} = -2.83.$$

This is a two-tailed test with α = .01. As a result, the critical values are ± 2.575. Since -2.83 is in the critical region, we reject H_0. We conclude that the evidence supports the claim of a change in reaction time.

11-23 Since n is greater than 25 we use the normal approximation. Our claim is that the majority favor the bill. Thus

H_0 : No more than half favor the bill

H_1 : A majority favor the bill

Then with x = 22, we have

$$z = \frac{(x + .5) - (n/2)}{\sqrt{n}/2} = \frac{22.5 - 25}{\sqrt{50}/2} = -.71.$$

This is a one-tailed z test, and hence α = .10, and the critical value is −1.28. Since our observed z value is more than -1.28, we would find no reason based upon the data to believe that more than half of the voters favored the bill.

<u>SOLUTIONS TO 11-2 B EXERCISES</u>

11-25 For the data n = 15. Using table A-8, the critical value for α = .05 is k = 3. We place the data in ascending order:

1 2 3 6 7 8 8 9 10 11 14 17 23 25 30.

Then $X_{k+1} = X_4 = 6$ (the 4th lowest number), and

$X_{n-k} = X_{15-3} = X_{12} = 17.$

The 95% confidence interval is $6 < M < 17$.

11-27 Since the number of negative results is 50, then we have more than 25 subjects. That is, n > 25. As a result, our statistic is

$$z = \frac{(x + .5) - (n/2)}{\sqrt{n}/2}.$$

As a one-tailed test with α = .01, our critical value of z = 2.33. Then we have

$$2.33 = \frac{50.5 - (n/2)}{\sqrt{n}/2} .$$

Then

$$2.33(\sqrt{n}/2) = 50.5 - (n/2)$$

$$2.33\sqrt{n} = 101 - n$$

$$n + 2.33\sqrt{n} - 101 = 0.$$

Using the quadratic equation

$$\sqrt{n} = \frac{-2.33 + \sqrt{(2.33)^2 + 404}}{2} = 8.95.$$

n = 80 when rounded up.

SOLUTIONS TO SECTION 11.3 - WILCOXON SIGNED-RANKS TEST FOR TWO DEPENDENT SAMPLES

11-29 Rearranging we have

Data Value	5	8	10	12	15
Rank	1	2	3	4	5

11-31 Ordering the data we have

Data	50	100	150	200	400	600
Rank	1	2	3	4	5	6

11-33 In order we have

Data	6	8	8	9	12	20
Rank	1	2.5	2.5	4	5	6

11-35 We have

Data	12	13	13	13	14	15	16	16	18
Rank	1	3	3	3	5	6	7.5	7.5	9

11-37 a, b, and c. We form a table similar to Table 11-4.

Before	After	Difference	Ranks of Difference	Signed Ranks
103	100	3	3	3
98	105	-7	5	-5
112	114	-2	2	-2
94	98	-4	4	-4
118	119	-1	1	-1
99	99	0	-	-
90	100	-10	6	-6
101	116	-15	7	-7

The ranks of differences were found by ranking the absolute values of the non-zero differences. The absolute values and ranks are

Differences	1	2	3	4	7	10	15
Rank	1	2	3	4	5	6	7

d. The value of T is then the smaller of:

1. $|-5| + |-2| + |-4| + |-1| + |-6| + |-7| = 25$

2. $|3| = 3$

11-39 a, b, and c. Forming a table similar to Table 11-4

Before	After	Difference	Ranks of Differences	Signed Ranks
83	82	1	1.5	1.5
76	77	-1	1.5	-1.5
91	89	2	3	3
59	62	-3	4	-4
62	68	-6	6	-6
75	70	5	5	5
80	90	-10	7	-7
66	86	-20	8	-8
73	73	0	--	--

The ranks of the differences were found by ranking the absolute values of the non-zero differences. These were

Differences	1	1	2	3	5	6	10	20
Rank	1.5	1.5	3	4	5	6	7	8

d. The value of T is the smaller of

1. $1.5 + 3 + 5 = 9.5$

2. $1.5 + 4 + 6 + 7 + 8 = 26.5$

Then $T = 9.5$.

11-41 a. Using $\alpha = .05$ and a two-tailed test, the critical value from Table A-8 is 25. Since $T = 24$ is less than that, we would reject H_0.

b. We have the same critical value as in a above. Since $T = 25$ is less than or equal to the critical value of 25, we would reject H_0.

11-43 a. Using $\alpha = .05$ and a two tailed test, the critical value from table A-8 is 35. Since $T = 25$ is less than that, we would reject the null hypothesis.

b. The critical value here is 8. Since $T = 8$ is less than or equal to the critical value, we would reject the null hypothesis.

11-45 Since the two tests were administered to the same individuals, we have paired data. Forming our table for analysis we have

Test A	Test B	Difference	Ranks of Differences	Signed Ranks
100	106	-6	5	-5
111	112	-1	1	-1
93	95	-2	2.5	-2.5
92	90	2	2.5	2.5
99	107	-8	6	-6
85	100	-15	9	-9
117	126	-9	7	-7
110	105	5	4	4
98	110	-12	8	-8

We compute T as the smaller of 2.5 + 4 = 6.5 and 5 + 1 + 2.5 + 6 + 8 + 7 + 8. Then T = 6.5. Using a two-tailed test with n = 9 and α = .05, the critical value from Table A-8 is 6. Since our observed T = 6.5, we fail to reject H_0. The data does support equality of the tests.

11-47 We form the table for the analysis.

Sitting	Supine	Difference	Ranks of Differences	Signed Ranks
4.66	4.63	.03	1	1
5.70	6.34	-.64	10	-10
5.37	5.72	-.35	5	-5
3.34	3.23	.11	2	2
3.77	3.60	.17	3	3
7.43	6.96	.47	8	8
4.15	3.66	.49	9	9
6.21	5.81	.40	6	6
5.90	5.61	.29	4	4
5.77	5.33	.44	7	7

We compute T as the smaller of 10 + 5 = 15 and 1 + 2 + 3 + 4 + 6 + 7 + 8 + 9 = 40. Using a two tailed test with n = 10 the critical value from table A-8 is 8 or less. Since T = 15 is not in the critical region, we fail to reject the null hypothesis that the two samples came from the same distribution.

11-49 Because we tested the two eyes of the same individuals, the data is paired. We form the table for analysis.

Right Eye	Left Eye	Difference	Ranks of Differences	Signed Ranks
14.7	15.2	-.5	3	-3
16.3	16.7	-.4	2	-2
12.4	12.6	-.2	1	-1
8.1	10.4	-2.3	4.5	-4.5
21.6	24.1	-2.5	6	-6
13.9	17.2	-3.3	8	-8
14.2	11.9	2.3	4.5	4.5
15.8	18.4	-2.6	7	-7

We use a two-tailed test that perception does not differ between the two eyes. Then n = 8. Clearly T = 4.5 since that is the smaller sum. The critical value from Table A-8 is 4. Since T is more than that, we fail to reject H_0. The data supports the claim of equal perception.

11-51 Testing was on the same students, so the data is paired. Forming our table we have

Classroom	Recess	Difference	Ranks of Differences	Signed Ranks
7.2	8.7	-1.5	10.5	-10.5
7.8	7.1	.7	4	4
6.0	5.8	.2	2	2
5.1	4.2	.9	6	6
3.9	4.0	-.1	1	-1
4.7	3.6	1.1	7	7
8.2	7.0	1.2	8	8
9.1	8.7	.4	3	3
8.7	7.9	.8	5	5
4.1	2.5	1.6	12	12
3.8	2.4	1.4	9	9
6.7	5.0	1.7	13	13
5.8	4.0	1.8	14	14
4.9	3.4	1.5	10.5	10.5

Adding the corresponding signed ranks, we have T = 11.5. We have n = 14 and α = .05. The critical value of T is 21. Since our value of 11.5 is less than 21, we reject H_0. The data supports the claim that anxiety levels differed between classroom and recess.

11-53 We are testing the claim that the two tests produce the same results of voters. We form a table of scores and ranks.

Test A	Test B	Differences	Ranks Of Differences	Signed Ranks
237	217	20	7	7
215	190	25	9	9
312	307	5	3	3
190	192	-1	1	-1
217	220	-3	2	-2
250	233	17	6	6
341	314	27	10	10
380	367	13	5	5
270	249	21	8	8
245	238	7	4	4

There are 10 scores which are ranked. Then T = 3 as the smaller of the signed ranks.
H_0 : The two tests yield the same results
H_1: The two tests yield different results
Using a two tailed test with α = .05, the critical value is 8. Since T = 3 is in the critical region, we reject the null hypothesis in favor of the conclusion that the two tests yield different results.

11-55 This problem is solved in a manner similar to 11-53.

SOLUTIONS TO 11-3 B EXERCISES

11-57 a. Our hypothesis is that both systems are the same. Then,
H_0: The two systems are the same
H_1: The systems are not the same
We have that n = 39. Since this exceeds the scope of the table A-8, we must use formula 11-1. The sum of the negative ranks will be smaller than that of the positive. Then, T = 2.5+4.5+6+7+10+21.5 =51.5. From formula 11-1 we have

$$z = \frac{T - n(n + 1)/4}{\sqrt{\frac{n(n + 1)(2n + 2)}{24}}} = \frac{51.5 - (36)(37)/4}{\sqrt{\frac{(36)(37)(74)}{24}}} = -4.42.$$

This is a two tailed test with α = .05. From the table for the normal distribution the critical values are ±2.575. The observed value of z is in the critical region and so we reject the null hypothesis that the systems are the same. The data supports a conclusion of nonequality of the two checkout systems.

b. Solved like part a except there are no ties so n = 39.

11-59 There are two critical values which are

$$T = \frac{n(n + 1)}{4} \pm z(\alpha/2) \sqrt{\frac{n(n + 1)(2n + 2)}{24}}$$

$$= \frac{100(101)}{4} \pm (1.96)(291) = 1955 \quad \text{or } 3096.$$

The smaller is 1955 which becomes the critical value.

SOLUTIONS TO SECTION 11-4 WILCOXON RANK-SUM TEST FOR TWO INDEPENDENT SAMPLES

11-61 Our hypothesis is that the performance is the same.
H_0: The two sets of scores are the same
H_1: The scores are different
The table with ranks in parentheses is as follows.

New York	California
520 (12)	508 (10)
490 (9)	563 (17)
571 (18)	385 (1)
398 (3)	617 (21)
602 (20)	704 (23)
475 (7)	401 (4)
557 (16)	409 (6)
621 (22)	527 (14)
737 (24)	393 (2)
403 (5)	478 (8)
511 (11)	521 (13)
598 (19)	536 (15)

If we sum the ranks for the California tests, we have R = 134. We compute μ and σ as follows.

$$\mu = \frac{n_1(n_1 + n_2 + 1)}{2} = \frac{12(12 + 12 + 1)}{2} = 150$$

$$\sigma = \sqrt{\frac{n_1 n_2(n_1 + n_2 + 1)}{12}} = \sqrt{300} = 17.32.$$

Then

$$z = \frac{R - \mu}{\sigma} = \frac{134 - 150}{17.32} = -.924.$$

The critical values for z with α = .05 are ±1.96. Since the observed z is not in the critical region, we fail to reject the null hypothesis of equality of the two samples.

11-63 Our hypothesis is that
H_0: Ratings by the two sexes are the same
H_1: Ratings are dependent upon sex
The table of ratings and ranks is as follows.

Males		Females	
27	(4)	21	(3)
36	(6)	43	(11.5)
42	(10)	38	(7)
57	(13)	40	(8.5)
88	(22)	40	(8.5)
92	(23)	60	(15)
60	(15)	87	(21)
60	(15)	72	(17)
43	(11.5)	73	(18)
29	(5)	10	(1)
76	(19)	12	(2)
79	(20)		

We compute z using R = 112.5, the sum of ranks for females. If R was computed using rankings of males, different values of the mean and standard deviation would be computed. Then $n_1 = 11$ and $n_2 = 12$.

$$\mu = \frac{n_1(n_1 + n_2 + 1)}{2} = \frac{11(11 + 12 + 1)}{2} = 132$$

$$\sigma = \sqrt{\frac{n_1 n_2(n_1 + n_2 + 1)}{12}} = \sqrt{264} = 16.25.$$

Then

$$z = \frac{R - \mu}{\sigma} = \frac{112.5 - 132}{16.25} = -1.20.$$

The critical values for z at $\alpha = .05$ are ± 1.96. Since the observed z is not in the critical region, we fail to reject the null hypothesis of equality of opinion of the two sexes.

11-65 Our hypothesis is

H_0: The two rail routes are equally fast

H_1: There is a difference in speed

The table of rates and ranks is as follows.

Route A		Route B	
98	(12)	96	(10)
102	(15.5)	132	(7)
83	(3)	121	(26)
117	(25)	87	(4)
128	(27)	106	(18)
92	(6)	102	(15.5)
112	(21)	116	(24)
108	(19.5)	95	(8.5)
108	(19.5)	99	(13)
100	(14)	76	(2)
93	(7)	97	(11)
72	(1)	104	(17)
95	(8.5)	115	(23)
91	(5)	114	(22)

We sum the ranks for route A to get R = 184. Then $n_1 = 14$ and $n_2 = 14$. We compute z as

$$\mu = \frac{n_1(n_1 + n_2 + 1)}{2} = \frac{14(14 + 14 + 1)}{2} = 203$$

$$\sigma = \sqrt{\frac{n_1 n_2(n_1 + n_2 + 1)}{12}} = \sqrt{473.67} = 21.76.$$

Then

$$z = \frac{R - \mu}{\sigma} = \frac{184 - 203}{21.76} = -.87.$$

The critical values for z at $\alpha = .05$ are $\pm$ 1.96. Since the test value of z is not in the critical region, we fail to reject the null hypothesis of equality of the two routes.

11-67 This problem is solved in a manner similar to 11-65.

11-69 Our hypothesis is
H_0: The attitudes of men and women are the same
H_1: The attitudes are different.
The table of ranks and scores are as follows.

Men		Women	
67	(16)	20	(5)
72	(18.5)	40	(10)
48	(12)	37	(9)
30	(6)	42	(11)
92	(25)	51	(13)
15	(3.5)	15	(3.5)
5	(1)	68	(17)
87	(23)	35	(8)
91	(24)	12	(2)
54	(14)	31	(7)
66	(15)	85	(22)
72	(18.5)		
98	(27)		
97	(26)		
75	(21)		
74	(20)		

Summing the ranks for the men, we have R = 270.5. Then n_1 = 11 and n_2 = 16. We compute z as

$$\mu = \frac{n_1(n_1 + n_2 + 1)}{2} = \frac{16(11 + 16 + 1)}{2} = 224$$

$$\sigma = \sqrt{\frac{n_1 n_2(n_1 + n_2 + 1)}{12}} = \sqrt{410.67} = 20.26.$$

Then

$$z = \frac{R - \mu}{\sigma} = \frac{270.5 - 224}{20.26} = 2.30.$$

The critical points for a z test at α = .05 are ±1.96. Since the test z is in the critical region, we reject the hypothesis of equality of opinion.

11-71 This problem is solved in a manner similar to 11-69.

11-73 The null hypothesis is that the two sets of scores come from the same population. We form the table of scores and their ranks.

Experimental		Control	
60.83	(8)	122.80	(35)
73.46	(16)	118.43	(29)
69.95	(11)	121.70	(33)
72.14	(15)	104.06	(26)
80.03	(21)	70.02	(13)
117.80	(28)	54.22	(5)
34.26	(2)	70.70	(13)
21.37	(1)	94.23	(24)
57.29	(6)	119.89	(31)
76.59	(20)	118.58	(30)
44.71	(4)	99.08	(25)
82.25	(22)	111.26	(27)
59.78	(8)	138.27	(36)

64.05 (10)	74.61 (18)
74.27 (17)	120.76 (32)
75.38 (19)	121.67 (34)
59.77 (7)	
92.72 (23)	
44.09 (3)	
66.87 (11)	

We will let R denote the sum of the ranks of the experimental group. Then R = 252. The values for μ and σ are

$$\mu = \frac{n_1(n_1 + n_2 + 1)}{2} = \frac{20(20 + 16 + 1)}{2} = 370$$

$$\sigma = \sqrt{\frac{n_1 n_2(n_1 + n_2 + 1)}{12}} = 31.4.$$

Then,

$$z = \frac{R - \mu}{\sigma} = \frac{254 - 370}{31.4} = -3.69.$$

The critical values of z for a two tailed test with α = .05 are ±1.96. Since the observed value of z is in the critical region, we reject the null hypothesis. The data supports a conclusion of inequality of the two sets of scores.

11-75 This problem is solved in the same manner as 11-73.

SOLUTIONS TO 11-4 B EXERCISES

11-77 a. The sum of the ranks of group A is R = 120. We compute z as

$$\mu = \frac{n_1(n_1 + n_2 + 1)}{2} = \frac{15(15 + 15 + 1)}{2} = 232.5$$

$$\sigma = \sqrt{\frac{n_1 n_2(n_1 + n_2 + 1)}{12}} = \sqrt{581.25} = 24.1.$$

Then

$$z = \frac{R - \mu}{\sigma} = \frac{120 - 232.5}{24.10} = -4.67.$$

At α = .05, the critical values are ±1.96. Since the observed z is in the critical region, we reject the hypothesis of equality.

b. For B the sum of ranks is 240. We compute z using the μ and σ computed in a above.

$$z = \frac{R - \mu}{\sigma} = \frac{240 - 232.5}{24.10} = .311.$$

This z is not in the critical region at α = .05. As a result we fail to reject the hypothesis of equality.

c. The ranks in a. were not compatible while the ranks of b. were.

d. There would be no change except that the group considered as being ranked higher would change.

e. If the observation ranked number 1 had a 30 ranking instead, then the sum R would have been increased by 29. Then R = 149. Then z would have become

$$z = \frac{149 - 232.5}{24.10} = -3.46.$$

We would have still rejected H_0 as was done in a.

11-79 a. We form our table of ranks and rank sums.

Rank 1	2	3	4	Rank Sum for A
A	A	B	B	3
A	B	A	B	4
A	B	B	A	5
B	A	A	B	5
B	A	B	A	6
B	B	A	A	7

b. If all sequences are equally likely, the table of rank sums and probabilities is given below.

Rank	Probability
3	1/6
4	1/6
5	2/6
6	1/6
7	1/6

c. No, since the minimum rank sum probability is 1/6 = .1666. Therefore no rank sum has probability less than .10.

SOLUTIONS TO SECTION 11-5 KRUSKAL-WALLIS TEST

11-81 The ranks of the three treatments are listed. The sums of the three ranks are $R_1 = 40$, $R_2 = 40$ and $R_3 = 40$. We compute H as

$$H = \frac{12}{N(N+1)}\left[\frac{R_1^2}{n_1} + \frac{R_2^2}{n_2} + \frac{R_3^2}{n_3}\right] - 3(N+1)$$

$$= \frac{12}{15(16)}\left[\frac{1600}{5} + \frac{1600}{5} + \frac{1600}{5}\right] - 3(16) = 0.$$

11-83 Our hypothesis is

H_0: The samples are from identical populations

H_1: The samples are not from identical populations

Adding the ranks as given in the problem we have

$R_1 = 29.5$ for the stable group

$R_2 = 51.5$ for the divorced group

$R_3 = 39$ for the transition group.

Then computing H we have

$$H = \frac{12}{N(N+1)}\left[\frac{R_1^2}{n_1} + \frac{R_2^2}{n_2} + \frac{R_3^2}{n_3}\right] - 3(N+1)$$

$$= \frac{12}{15(16)}\left[\frac{870.25}{5} + \frac{2652.25}{5} + \frac{1521}{5}\right] - 3(16) = 2.435.$$

This is a χ^2 random variable with 2 degrees of freedom (there are three groups). Using $\alpha = .05$, the critical value is 5.991. Since the observed value of $H = 2.435$ does not exceed this, we fail to reject the hypothesis that the samples are from identical populations.

11-85 Our hypothesis is

H_0: The three groups scored the same

H_1: The three groups were not the same

We give the table of scores and ranks as follows.

Good Record		Fair Record		Poor Record	
90	(16)	80	(12.5)	60	(4.5)
86	(14)	70	(9.5)	60	(4.5)
88	(15)	61	(6)	55	(3)
93	(17)	52	(2)	62	(7)
80	(12.5)	73	(11)	50	(1)
96	(18)	65	(8)	70	(9.5)

Summing the ranks we have
R_1 = 92.5 for good record students
R_2 = 49 for fair record students
R_3 = 29.5 for poor record students.
Computing H we have

$$H = \frac{12}{N(N+1)}\left[\frac{R_1^2}{n_1} + \frac{R_2^2}{n_2} + \frac{R_3^2}{n_3}\right] - 3(N+1)$$

$$= \frac{12}{18(19)}\left[\frac{8556}{6} + \frac{2401}{6} + \frac{870}{6}\right] - 3(19) = 12.16.$$

As in 11-71, the critical value for χ^2 is 5.991. Since the observed value of H exceeds this, we reject the null hypothesis and conclude that there is evidence of differences in the populations.

11-87 The solution here proceeds like 11-73.

11-89 The null hypothesis here is that the data on living areas from the three zones all comes from the same population. The data and ranks are presented below.

Zone 1	Zone 4	Zone 7
20 (19)	18 (14.5)	21 (21)
18 (14.5)	20 (19)	16 (9)
27 (24)	19 (17)	14 (6)
17 (11)	12 (3.5)	23 (22)
18 (14.5)	9 (1)	17 (11)
13 (5)	18 (14.5)	17 (11)
24 (23)	28 (25)	20 (19)
15 (17.5)	10 (2)	
	15 (17.5)	
	12 (3.5)	

The sums of ranks are for zone 1 R_1 = 118.5, for zone 4 R_2 = 107.5, and for zone 7 R_3= 99. Computing H we have

$$H = \frac{12}{N(N+1)}\left[\frac{R_1^2}{n_1} + \frac{R_2^2}{n_2} + \frac{R_3^2}{n_3}\right] - 3(N+1)$$

$$= \frac{12}{25(26)}\left[\frac{14042}{8} + \frac{11556}{10} + \frac{9801}{7}\right] - 3(26)$$

$$= 1.59$$

There are 2 degrees of freedom for this χ^2 distribution. The critical value for α = .05 is 5.991. Since H does not exceed this, we fail to reject the null hypothesis of equal living areas.

11-91 This exercise is solved in a manner similar to 11-89.

11-93 Our hypothesis is
H_0: The 5 methods yield equivalent results
H_1: The 5 methods do not yield equivalent results
The table of scores and their ranks is given below.

Trad	Prog	Audio	AV	Visual
76.2 (19)	85.2 (27)	67.3 (9)	75.8 (18)	50.5 (2)
78.3 (22)	74.3 (17)	60.1 (5)	81.6 (25)	70.2 (13)
85.1 (26)	76.5 (20)	55.4 (3)	90.3 (30)	88.8 (29)
63.7 (7)	80.3 (24)	72.3 (15)	78.0 (22)	67.1 (8)
91.6 (31)	67.4 (10)	40. (1)	67.8 (11)	77.7 (21)
87.2 (28)	67.9 (12)		57.6 (4)	73.9 (16)
	72.1 (14)			
	60.4 (6)			

The sum of the ranks is
R_1 = 135 for traditional
R_2 = 130 for programmed
R_3 = 33 for audio
R_4 = 110 for audio-visual
R_5 = 89 for visual.
The statistic H is given by

$$H = \frac{12}{N(N+1)}\left[\frac{R_1^2}{n_1} + \frac{R_2^2}{n_2} + \frac{R_3^2}{n_3} + \frac{R_4^2}{n_4} + \frac{R_5^2}{n_5}\right] - 3(N + 1) =$$
$$= \frac{12}{31(32)}\left[\frac{18225}{6} + \frac{16900}{8} + \frac{1089}{5} + \frac{12100}{6} + \frac{7521}{6}\right] - 3(32) = 8.76.$$

The critical value for the χ^2 variable with 4 degrees of freedom and α = .05 is 9.4877. The observed value of H = 8.49 fails to exceed this level. As a result, we fail to reject the hypothesis of equivalence of the teaching methods.

11-95 This problem is solved in a manner similar to 11-93.

SOLUTIONS TO 11-5 B EXERCISES

11-97 In formula 11-3 we would have 8 samples of 6. Therefore N = 48. Also the n_i are all 6. The formula becomes

$$H = \frac{12}{48(49)}\left[\frac{R_1^2}{6} + \frac{R_2^2}{6} + \frac{R_3^2}{6} + \frac{R_4^2}{6} + \frac{R_5^2}{6} + \frac{R_6^2}{6}\right] - 3(N + 1)$$
$$= \frac{1}{1176}\left[\sum R_i^2\right] - 147$$

11-99 Referring to the data of the example, there are ties at 147000 and 160000, each having t = 2 ties. Then we get $T = t^3 - t = 2^3 - 2 = 6$ for each tie. As a result, $\sum T = 12$. We divide the value of H by

$$1 - \frac{\sum T}{N^3 - N} = 1 - \frac{12}{25^3 - 25} = .999.$$

The resultant value of H is 1.138/.999 = 1.139.

SOLUTIONS TO SECTION 11-6 RANK CORRELATION

11-101 a. From the table the critical values are ±.450.

b. Using the formula, r_s is given by

$$r_s = \pm\frac{z}{\sqrt{n-1}} = \pm\frac{1.96}{\sqrt{49}} = \pm.28.$$

c. From the formula we have

$$r_s = \pm \frac{z}{\sqrt{n-1}} = \pm \frac{2.33}{\sqrt{39}} = \pm.373.$$

11-103 The table of scores and ranks is as follows.

Scores x	Scores y	Ranks x	Ranks y
2	30	1	5
5	25	2	4
8	20	4	3
6	15	3	2
10	5	5	1

11-105 We form a table of scores, ranks and differences as follows.

Scores x	Scores y	Ranks x	Ranks y	d	d^2
1	4	1	4	-3	9
2	2	2	2	0	0
3	3	3	3	0	0
4	1	4	1	3	9

Then $\sum d^2 = 18$ and r_s is given by

$$r_s = 1 - \frac{6\sum d^2}{n(n^2 - 1)} = 1 - \frac{6(18)}{4(4^2 - 1)} = -.8.$$

11-107 The table of scores, ranks and differences is given below.

Score x	Score y	Ranks x	Ranks y	d	d^2
63	43	2	5	-3	9
68	44	3	6	-3	9
71	39	5	4	1	1
55	30	1	3	-2	4
70	28	4	2	2	4
75	20	6	1	5	25

Then $\sum d^2 = 52$ and the rank correlation is

$$r_s = 1 - \frac{6\sum d^2}{n(n^2 - 1)} = 1 - \frac{6(52)}{6(6^2 - 1)} = -.486.$$

11-109 a. Since the ranks are given, we need only list the differences and their squares. Forming salary - SAT score ranks, we have

d	-3	-1	2	-2	-1	0	1	4
d^2	9	1	4	4	1	0	1	16.

Then $\sum d^2 = 36$. Then the rank correlation is

$$r_s = 1 - \frac{6\sum d^2}{n(n^2 - 1)} = 1 - \frac{6(36)}{8(8^2-1)} = .571.$$

b. The critical values from table A-9 are $\pm.738$

c. The data does not support a conclusion of correlation because the correlation from the data is between the critical values.

11-111 a. As in 11-109, the ranks are given so we only list differences and their squares. Forming SAT - aid ranks, we have

d	2	2	-2	3	-1	2	-3	-3
d^2	4	4	4	9	1	4	9	9

Therefore, $\sum d^2 = 44$. Then the rank correlation is

$$r_s = 1 - \frac{6\sum d^2}{n(n^2 - 1)} = 1 - \frac{6(44)}{8(8^2-1)} = .476.$$

b. The critical values from table A-9 are ±.738.
c. Since .476 is not in the critical region, the data does not support a claim of correlation.

11-113 a. We rank the data

Exports	1	2	3	6	4.5	4.5
Imports	1	2	3	4	5	6
d	0	0	0	2	-.5	-1.5
d^2	0	0	0	4	.25	2.25

Then $\sum d^2 = .5$. Computing the rank correlation, we have

$$r_s = 1 - \frac{6\sum d^2}{n(n^2 - 1)} = 1 - \frac{6(6.5)}{6(6^2-1)} = .814.$$

b. The critical values from the table are ±.886.

c. Since .814 is between the critical values, we find that the data supports a claim of positive correlation.

11-115 a. We list the ranks as follows.

Load	1	2	3	4	5	6	7	8
Length	1	2	3	4	5	6	7	8
d	0	0	0	0	0	0	0	0

Since all differences are 0, then $\sum d^2 = 0$. Then

$$r_s = 1 - \frac{6\sum d^2}{n(n^2 - 1)} = 1 - \frac{6(0)}{8(8^2-1)} = 1.$$

b. The critical values from the table are ±.738 for $\alpha = .05$.
c. Since 1. is in the critical region, we conclude that the data supports a conclusion of positive correlation.

11-117 a. Since the ranks are given, we need only list the differences and their squares. Forming Allen - Zeller, we will have

d 2, -1, 0, -1, 2, 0, 0, -1, -3, 2
d^2 4, 1, 0, 1, 4, 0, 0, 1, 9, 4

We have $\sum d^2 = 22$. Then we compute r_s as

$$r_s = 1 - \frac{6\sum d^2}{n(n^2 - 1)} = 1 - \frac{6(24)}{10(10^2-1)} = .855.$$

b. From the table the critical values are ±.648.

c. Since the observed rank correlation exceeds the critical value of .648, we conclude that the evidence supports positive correlation between the two variables.

11-119 a. Since the ranks are given, we need only list the differences and their squares. Forming office minus hotel, we have

d 0, 1, -1, -1, 0, 3, 0, 0, 7, -1, -1, -1, -1, -6, 1
d^2 0, 1, 1, 1, 0, 9, 0, 0, 49, 1, 1, 1, 1, 36, 1

and $\sum d^2 = 60$. Then

$$r_s = 1 - \frac{6\sum d^2}{n(n^2 - 1)} = 1 - \frac{6(102)}{15(15^2-1)} = .818.$$

b. Using the table, the critical values for r are ±.525.

c. Since the observed rank correlation exceeds the critical value, we would reject the null hypothesis of no correlation. We conclude that the evidence supports positive correlation between the two variables.

11-121 This problem is solved like 11-119.

11-123 This problem is solved in the same manner as 11-119.

SOLUTIONS TO 11-6 B EXERCISES

11-125 a. There are 6 different ways the three could have been ranked. They are

1	2	3	4	5	6
1	1	2	2	3	3
2	3	1	3	1	2
3	2	3	1	2	1

b. We compute the values of $\sum d^2$ as compared with a ranking of 1,2,3. In each case the rank correlation is also computed.

1) $\sum d^2 = 0$ and so $r_s = 1$
2) $\sum d^2 = 2$ and so $r_s = .5$
3) $\sum d^2 = 2$ and so $r_s = .5$
4) $\sum d^2 = 6$ and so $r_s = -.5$
5) $\sum d^2 = 6$ and so $r_s = -.5$
6) $\sum d^2 = 8$ and so $r_s = -1$

c. Only one sample value of r_s is more than .9. As a result, since there are 6 equally likely observations, the probability is 1/6.

SOLUTIONS TO SECTION 11-7 RUNS TEST FOR RANDOMNESS

11-129 Let n_1 be the number of As and n_2 be the number of Bs. Then $n_1 = 5$ and $n_2 = 3$. There is a total of G = 3 runs. From table A-10 the critical values for $\alpha = .05$ are 1 and 8. Since G is in the interval from 2 to 7, we would fail to reject a hypothesis of randomness of the data.

11-131 Here n_1 is the number of As = 4 and n_2 is the number of Bs = 6. There are G = two runs. From table A-10 with $\alpha = .05$ the critical values are 2 and 9. Since G is in the critical region of less than or equal to 2, we reject the hypothesis of randomness of the sample data.

11-133 We have $n_1 = 7$ and $n_2 = 2$. There are G = 2 runs. From the table the critical values are 1 and 6. Since G = 2 is not in the critical region, we fail to reject a hypothesis of randomness of the data.

11-135 We have $n_1 = 10$ and $n_2 = 10$. There are G = 9 runs. The critical values for G are 6 and 16. Since G = 9 runs is not in the critical region, we fail to reject the hypothesis of randomness of the data.

11-137 a. The mean is 11.

b. Stating the observations as A = above the mean and B = below, the observations are

B B B B B A B A A A.

c. From part b, the parameters are $n_1 = 4$, $n_2 = 6$, and G = 4 runs.

d. From the table the critical values are 2 and 9.

e. Since $G = 4$ runs is not in the critical region, we fail to reject the hypothesis of randomness.

11-139 a. The mean is 129/14 = 9.21.

b. Stating the observations as A = above the mean and B = below, we have
A A A A A A B B B B B B B B.

c. Using $n_1 = 6$ and $n_2 = 8$, there are $G = 2$ runs.

d. The critical values from the table are 3 and 12.

e. Since $G = 2$ runs is in the critical region, we would reject the hypothesis of randomness of the data.

11-141 a. To find the median we order the data
3, 7, 7, 8, 9, 10, 12, 16, 18, 20.
The median is (9 + 10)/2 = 9.5.

b. With the observations as A = above the median and B = below, the ordering is
B B B B B A A A A A.

c. The parameters are $n_1 = 5$ and $n_2 = 5$. There are $G = 2$ runs.

d. From table A-10 the critical values for G are 2 and 10.

e. Since $G = 2$ runs falls in the critical region, we reject the hypothesis of randomness of the data.

11-143 a. Ordering the data we have
3, 3, 4, 5, 8, 8, 8, 9, 9, 9, 10, 13, 14, 15, 16, 18, 20.
The median is 9 for this data.

b. We delete the three observations which equal the median. For A = above the median and B = below, the data is
B B B B B B B A A A A A A A.

c. The parameters are $n_1 = 7$ and $n_2 = 7$. There are $G = 2$ runs.

d. From table A-10 the critical values are 3 and 13.

e. The number of runs is in the critical region. Therefore we reject the hypothesis of randomness of the data.

11-145 The null hypothesis is
H_0: The polling of males and females is random in the sample
H_1: The polling is not random in the sample.
There are 27 people in the poll of which 9 are female. Therefore, $n_1 = 9$ and $n_2 = 18$. Since $G = 4$ falls in the critical region, we reject the null hypothesis of randomness. The data supports a conclusion of nonrandomness.

11-147 The hypotheses are
H_0: The occurrence of defective items is random
H_1: The occurrence of defective items is not random.

There is a total of 45 items in the sample. Of these 15 were defective. We set $n_1 = 30$ and $n_2 = 15$. Also G = 14 runs. Critical values for these numbers can not be found in table A-10. Consequently we use the normal approximation.

$$\mu_G = \frac{2n_1n_2}{n_1 + n_2} + 1 = \frac{2(30)(15)}{30 + 15} + 1 = 21$$

$$\sigma_G = \sqrt{\frac{(2n_1n_2)(2n_1n_2 - n_1 - n_2)}{(n_1 + n_2)^2(n_1 + n_2 - 1)}}$$

$$= \sqrt{\frac{(2)(30)(15)(900 - 30 - 15)}{(15 + 30)^2(15 + 30 - 1)}} = 2.94.$$

We now calculate the z statistic.

$$z = \frac{G - \mu}{\sigma} = \frac{14 - 21}{2.94} = -2.38.$$

At the $\alpha = .05$ level, the critical values of z are -1.96 and 1.96. The test value of -2.38 is in the critical region. Therefore we reject the null hypothesis of randomness of the data.

11-149 To determine the median, we must put the data in order.
598, 617, 630, 655, 676, 720, 731, 812, 856, 899, 903, 905, 971, 973, 1067, 1456. There are 16 data points, so the median is (812+856)/2 = 834. Returning to the data and indicating A as above the median with B as below, we have the following order of observations.
B, B, B, B, B, B, B, B, A, A, A, A, A, A, A, A.
There are G = 2 runs. Also, $n_1 = 8$ and $n_2 = 8$. From the table the critical values are 4 and 14. Since G = 2 is in the critical region, we reject the hypothesis of randomness of the data.

11-151 Our hypothesis is

H_0 : The sequence is random with respect to age

H_1 : The data is not random

By categories the data is
yyyyyyoyooyyyyooooooooooooooyyyooooyyoyoooooo.
There is a total of 45 observations. There are G = 12 runs. Also $n_1 = 17$ younger children and $n_2 = 28$ older. We must use the formulas to compute the critical values.

$$\mu_G = \frac{2n_1n_2}{n_1 + n_2} + 1 = \frac{2(17)(28)}{17 + 28} + 1 = 22.2$$

$$\sigma_G = \sqrt{\frac{(2n_1n_2)(2n_1n_1 - n_1 - n_2)}{(n_1 + n_2)^2(n_1 + n_2 - 1)}}$$

$$\sqrt{\frac{(2)(17)(28)(952 - 17 - 28)}{(17 + 28)^2(17 + 28 - 1)}} = 3.11.$$

We compute z by

$$z = \frac{G - \mu}{\sigma} = \frac{12 - 22.2}{3.11} = -3.27.$$

The critical z values for $\alpha = .05$ are -1.96 and + 1.96. Since the test z is in the critical region, we reject the hypothesis of randomness of the data.

11-153 This problem is solved in a manner similar to 11-151.

11-155 This problem is solved in a similar manner to 11-151.

11-157 In this case both n_1 and n_2 are 30. We compute μ and σ as

$$\mu_G = \frac{2n_1n_2}{n_1 + n_2} + 1 = \frac{2(30)(30)}{30 + 30} + 1 = 31.$$

$$\sigma_G = \sqrt{\frac{(2n_1n_2)(2n_1n_1 - n_1 - n_2)}{(n_1 + n_2)^2(n_1 + n_2 - 1)}}$$

$$= \sqrt{\frac{(2)(30)(30)(1800 - 30 - 30)}{(60)^2(30 + 30 - 1)}} = 3.84.$$

Then,

$$z = \frac{G - \mu}{\sigma} = \frac{8 - 31}{3.84} = -5.99.$$

Since this is less than the critical value of $z = -1.96$ for $\alpha = .05$, we reject the hypothesis of randomness of the data.

11-159 The median is the point above which is half of the data. Therefore $n_1 = 40$. As a result $n_2 = 40$. We are given that $G = 6$. From this point the solution follows that of 11-157.

SOLUTIONS TO 11-7 B EXERCISES

11-161 The minimum number of runs is 2: A, A, B, B. Therefore the minimum G = 2. The maximum number of runs is 4: A, B, A, B. From the table the critical values are 1 and 6. It is not possible for this data to create a number of runs which will fall in the critical region. Consequently, it is not possible to conclude from a runs test that 4 observations, 2 of each type, came in a nonrandom fashion.

SOLUTIONS TO REVIEW EXERCISES

11-165 The table of differences and their ranks is given below.

Before	After	Difference	Ranks of Differences	Signed Ranks
63	73	10	8	8
65	64	-1	1.5	-1.5
67	72	5	6	6
70	74	4	5	5
71	77	6	7	7
77	80	3	4	4
78	79	1	1.5	1.5
80	82	2	3	3
85	96	11	9	9
91	91	0		

T = 1.5 as the smaller of the two sums of ranks. In this case there are 9 ranks. The critical value from table A-8 is 6. Since T is in the critical region, we reject the null hypothesis of no change.

11-167 We will use the Wilcoxon rank-sum test. We form our table of scores and ranks in parentheses.

Kings County		Queens County	
76	(18)	67	(15)
52	(9.5)	48	(8)
39	(6)	52	(9.5)
27	(4)	40	(7)
88	(21)	53	(11)
73	(16)	91	(22)
75	(17)	30	(5)
92	(23)	20	(2)
99	(24)	18	(1)
83	(20)	23	(3)
79	(19)	61	(12)
		63	(14)
		62	(13)

For the test statistics we have
R_1 = 177.5 for ranks for Kings County
R_2 = 122.5 for ranks for Queens County
Then

$$\mu = \frac{n_1(n_1 + n_2 + 1)}{2} = \frac{11(11 + 13 + 1)}{2} = 137.5$$

σ = 71.26 from the formula. Then $z = \dfrac{177.5 - 137.5}{17.26} = 2.32$

The critical values for α = .05 are ±1.96. Since z exceeds the critical value, we conclude that there is significant differences between the sets of scores.

11-169 The table of data, ranks and differences is given below.

Data		Ranks		d=differences	d^2
Age	Income	Age	Income		
27	3600	1	8	-7	49
42	120	6	2	4	16
29	2800	2	7	-5	25
30	1200	3	5.5	-2.5	6.25
32	1200	4	5.5	-1.5	2.25
55	50	7	1	6	36
57	140	8	3	5	25
33	900	5	4	1	1

Then $\sum d^2$ = 160.5. Our test is H_0: there is no correlation and H_1: there is some correlation. We compute the sample statistics as

$$r_s = 1 - \frac{6(160.5)}{8(8^2 - 1)} = -.911.$$

Then the critical values for r_s are

$$r_s = \pm \frac{z}{\sqrt{n - 1}} = \pm \frac{1.96}{2.64} = \pm.74.$$

Since the observed correlation of -.911 is in the critical region, we reject the hypothesis of no correlation between age and income.

11-171 We form a table of scores and their ranks in parenthesis.

Kings County 76(18), 52(9.5), 39(6), 27(4), 88(21), 73(16), 75(17), 92(23), 99(24), 83(20), 79(19)

Queens County 67(15), 48(8), 52(9.5), 40(7), 53(11), 91(22), 30(5), 20(2), 18(1), 23(3), 61(12), 63(14), 62(13)

We sum the ranks for Queens County to get R = 122.5. Computing the statistics we have

$$\mu = \frac{n_1(n_1 + n_2 + 1)}{2} = \frac{11(25)}{2} = 137.5$$

$$\sigma = \sqrt{\frac{n_1 n_2(n_1 + n_2 + 1)}{12}} = 17.26$$

and $$z = \frac{R - \mu}{\sigma} = \frac{177.5 - 137.5}{17.26} = 2.32.$$

The critical values for z using $\alpha = .05$ are ±1.96. Since the test z exceeds 1.96, we reject the hypothesis of equivalence of the two sets of scores.

11-173 Our hypothesis is

H_0: The judges rate the same

H_1: The judges do not rate the same

Our table for the signed ranks test is as follows.

Judge A	Judge B	Difference	Ranks	Signed Ranks
6.3	7.1	-.8	4	-4
7.2	6.5	.7	2.5	2.5
6.6	8.2	-1.2	7	-7
8.5	8.6	-.1	1	-1
9.7	9.0	.7	2.5	2.5
7.0	6.1	.9	5	5
7.3	6.3	1.0	6	6
8.8	8.8	0.	-	-

Then T = the smaller of the two ranked sums, which is 12. There are n = 7 pairs of data points. From table A-9, the critical value is 2. Since the observed T = 12 is not less than or equal to the critical value we fail to reject the hypothesis of no differences between the two judges.

11-175 We rank the data as follows.

Judge A	Judge B	Rank A	Rank B	Difference	D^2
6.3	7.1	1	4	-3	9
7.2	6.5	4	3	1	1
6.6	8.2	2	5	-3	9
8.5	8.6	6	6	0	0
9.7	9.0	8	8	0	0
7.0	6.1	3	1	2	4
7.3	6.3	5	2	3	9
8.8	8.8	7	7	0	0

Then $\sum d^2 = 32$ and

$$r_s = 1 - \frac{6\sum d^2}{n(n^2-1)} = 1 - \frac{6(32)}{8(63)} = .619.$$

We test for significance of the correlation coefficient. The critical value is

$$r_s = \pm \frac{z}{\sqrt{n-1}} = \pm \frac{1.96}{2.65} = \pm .74.$$

Since the observed value of the correlation coefficient is not in the critical region, we fail to reject the null hypothesis of no correlation.

11-177 To test the claim of a median, we shall use the sign test. Our hypothesis is
H_0: The median is 38
H_1: The median is not 38
Using a + for above the median and - for below, there are 10 minus signs, 9 plus signs and 1 tie. Then n = 19. From the table the critical value is 4. Since the observed number of positive signs is 9 and thus not in the critical region, we fail to reject the null hypothesis that the median is 38.

11-179 To test equal production, we will use the Kruskal-Wallis test. The observations and associated ranks are given below.

Machine A	Machine B	Machine C
660(11)	590(9)	520(1)
690(14.5)	588(8)	572(6)
690(14.5)	560(3)	578(7)
672(12)	570(5)	553(2)
683(13)	592(10)	564(4)

For each machine the sample size is n = 5. Sums of the column ranks are R_1 = 65, R_2 = 35, and R_3 = 20. The test statistic is

$$H = \frac{12}{N(N+1)}\left[\frac{R^2_1}{n_1} + \frac{R^2_2}{n_2} + \frac{R^2_3}{n_3}\right] -3(N + 1)$$
$$= \frac{12}{15(16)}\left[\frac{4225}{5} + \frac{1225}{5} + \frac{400}{5}\right] -3(16) = 10.5.$$

The critical value for H is determined from the χ^2 distribution. There are 2 degrees of freedom. For α = .05 the critical value is 5.991. Since H exceeds the critical value, we reject the null hypothesis thaount.

11-181 The runs test is the test used for randomness. Letting n_1 be the number for presence and n_2 for absence, then n_1 = 5 and n_2 = 20. There are G = 5 runs. The critical value from the runs table is less than or equal to 5 or greater than or equal to 12. Since the value of G = 5 is in the critical region, we reject the null hypothesis of randomness.

11-183 We set up the table of scores and associated ranks as follows.

A	B	C
.11(17)	.08(11)	.04(2)
.10(15.5)	.09(13)	.04(2)
.09(13)	.07(9.5)	.05(5)
.09(13)	.07(9.5)	.05(5)
.10(15.5)	.06(7.5)	.06(7.5)
		.04(2)
		.05(5)

We sum the ranks and get R_1 = 74, R_2 = 50.5, and R_3 = 28.5. Then or the Kruskal-Wallis test we have

$$H = \frac{12}{N(N+1)}\left[\frac{R^2_1}{n_1} + \frac{R^2_2}{n_2} + \frac{R^2_3}{n_3}\right] - 3(N + 1)$$
$$= \frac{12}{17(18)}\left[\frac{5476}{5} + \frac{2550}{5} + \frac{812}{7}\right] - 3(18) = 13.5.$$

H has a χ^2 distribution with k-1 = 3-1 = 2 degrees of freedom. The critical value at α = .05 is 5.991. Since H exceeds that value. we reject the hypothesis of equality.

11-185 We use the runs test to test for randomness. Letting n_1 be the number of males and n_2 the number of females, then n_1 = 20 and n_2 = 10. There are G = 5 runs.

From table A-10 the critical region is less than or equal to 9 or more than or equal to 20. Since the observed number of runs is in the critical region, we reject the hypothesis of randomness.

11-187 Setting the data as before minus after, there is 1 + sign and 11 - signs with two ties. Using table A-7 with $\alpha = .05$ and $n = 12$, the critical two tailed region is less than or equal to 2. Since the lesser of 1 and 11 is in the critical region, we reject the null hypothesis that the tune-up has no effect on fuel consumption. The data supports the claim that fuel consumption improves.